Efficiency in Research, Develop
The Statistical Design and Anal
Experiments

Efficiency in Research, Development, and Production:

The Statistical Design and Analysis of Chemical Experiments

Leslie Davies
Department of Chemistry and Applied Chemistry,
University of Salford

ISBN 0-85186-137-7

A catalogue record for this book is available from the British Library

Published by The Royal Society of Chemistry,
Thomas Graham House, Cambridge CB4 4WF

Typeset by Vision Typesetting, Manchester
Printed by Henry Ling Ltd., at the Dorset Press, Dorchester

Preface

Chemists know very well that when they manufacture chemicals, irrespective of whether the yield is in grams or tons, there are numerous variables which may affect the outcome: time, temperature, concentrations and order of addition of reagents, nature and concentration of catalyst, source of materials, stirring rate, reactor design, ambient atmosphere, performance of operator, and so on. The analytical chemist is often concerned with many of these, and also with, for example, the source and storage of standard materials, sample selection, separation, and preparation, and the calibration, method of operation, and condition of analytical instruments. To examine all (or even most) of these systematically is prohibitively time-consuming and expensive unless a logical basis is found to minimize the necessary effort and yet maximize the probability that important variables will be identified, the magnitudes of their effects estimated, and optimum operating conditions developed. The techniques to achieve these objectives are those of experimental design.

Major chemical companies have long been aware that statistical methods have an important part to play in research and production; for example, ICI arranged with Oliver and Boyd Ltd for the publication of texts on statistical analysis[1] and experimental design.[2] Of these two aspects, *design is the more important*, for conclusions are easily obtained from a well-designed experiment, whereas even the best statistical analysis cannot extract much information from an ill-designed investigation. It is also the case that good design can lead to major improvements in productivity and profitability. It follows that, in this book, design is the principal subject of attention, although of course both aspects must go together.

Statistically-designed experiments are very much more economical of time and effort than the classical research approach of altering only one variable at a time. By systematically altering the values of one, two, or even several variables from one unit experiment to the next, designs known as factorial experiments yield good estimates of the effects of the variables, and also give additional important information which the classical method cannot (without a great deal of elaboration). This additional information measures the magnitudes of interactions, which are both common and important in chemistry. For example, a given change in reaction time would not be expected to have the same effect at different temperatures (there is a time–temperature interaction). Interaction effects are estimated, automatically and for no additional experimental effort, in factorial designs.

Various experimental design software packages are now available, but the essential thing for the chemist is to *understand the principles* of the methods, by having them explained and working out designs and results for oneself. In that way, their logic and applicability are appreciated, and the experimenter can learn to choose designs which suit his or her specific requirements. This book has been written with that objective in view, and introduces the reader to major techniques of design and analysis. The text includes illustrative examples (with model answers and discussions immediately following) and practice exercises with brief answers. The discussions are an important part of the text, and should be studied by the reader *after* attempting the examples. Readers requiring additional methods or more explanation will find the textbooks listed in the bibliography a rich source of information. Individual papers have been referred to where this seemed necessary or particularly appropriate.

REFERENCES

1. 'Statistical Methods in Research and Production', ed. O.L. Davies, Oliver and Boyd, Edinburgh and London, 1947.
2. 'Design and Analysis of Industrial Experiments', ed. O.L. Davies, Oliver and Boyd, Edinburgh, 1954.

Contents

Acknowledgements

Anyone who writes about the statistical design and analysis of experiments is deeply indebted to the late Sir Ronald Fisher, FRS, the originator of many of the fundamental ideas in the subject, to Dr Frank Yates, FRS, and Professor George Box, FRS, whose names appear often in this book, and to other distinguished statisticians, some of whom are also named in the text. My personal thanks also go to Professor E.G. Edwards, who made it possible for me, and numerous others, to gain our introduction to statistics from Professor Box and his (then) ICI colleagues in this institution long years ago; and to Mr S.T.R. Hancock who, before that, ensured that chemistry students like me learned some mathematics whether we liked it or not. My thanks also go to the Royal Society of Chemistry for encouraging me to write this book, to Mrs Mary Harris for her care and co-operation in preparing the typescript, and to Mr Alan Cubitt (of the Royal Society of Chemistry) and Vision Typesetting similarly, for their work in producing the final version.

Leslie Davies
Department of Chemistry and Applied Chemistry
University of Salford
January 1993

CHAPTER 1

Economy and Efficiency in Research and Development

1.1 AN INTRODUCTION TO TYPES OF EXPERIMENTAL DESIGN

Whenever we design and carry out experiments, obtain results, reach conclusions, and base recommendations on them, we want all of these to have validity. That is to say, we want them to be soundly based in theory and practice, and applicable to our purposes. We also want the whole process to occur with efficiency and despatch.

In order to begin to explain how experimental design and analysis can help us to do this is in a very economical and productive way, I shall risk seeming eccentric by starting with three quotations:

'Of every Quality, Comparison the only measure is, and judge, Opinion.'
John Donne, 1601

'Science . . . is as subjective and psychologically conditioned as any other . . . human behaviour.'
Albert Einstein, *c.* 1934

'Science cannot discover truth, but is an excellent means of discovering error. The residuum left after error has been eliminated is usually called scientific truth.'
Kenneth Boulding, 1969

This philosophizing may seem remote from everyday life in laboratory or plant. But it isn't: every scientific observation or measurement is a comparison. Every estimate we have of the effect of changing a system variable from one level to another is the result of such a comparison of two (or more) observations. We have to judge what such comparisons mean, and form a conclusion (give an opinion, at least to ourselves) as to what to do as a result. We have to do this as objectively as possible, trying to decide what is right and not what we (or the boss) might want. We have to reduce our errors and our uncertainties as much as possible (while not expecting that we can in fact eliminate them all).

As already mentioned, statistically designed experiments are highly efficient in that they give a fixed amount of information with much less effort than the classical one-variable-at-a-time approach, and many of them give additional information about interactions as a bonus. It will be shown later that significant interactions are an important clue in the search for optimum conditions, and substantial interactions mean that careful control will have to be exercised if a reproducible process is to ensue.

In all the investigations, the individual unit experiments are carried out in random order. As a consequence, the results as they emerge do not form any immediately obvious pattern: a good guard against premature jumping to conclusions – *i.e.* against subjectivity. We have a variety of significance tests, to help us to decide objectively on the likely meaning of the results. The method of analysis known as the analysis of variance automatically calculates estimates of experimental error if replicate experiments are done, in order to generate the final results; and the error magnitude can also suggest whether there are unidentified significant sources of variation to be looked for. A significance test is yet another comparison, testing the magnitude of a systematically occurring effect against the experimental error.

I shall use *factorial* and *fractional–factorial experiments* to illustrate the techniques alluded to above. Such designs have been described as 'the most powerful (statistical) techniques in technological research',[1] and well chosen fractional factorials are particularly economical in assessing multivariable systems. I also hope to show that the handling of the data is quite simple. What is essential is arithmetical accuracy and a willingness to accept and remember some rules. The other point to note is that the whole design (or a known fraction or fractions) must be worked through, to allow proper analysis (although techniques exist to estimate missing values).

The fundamental procedure in designing, carrying out, and analysing an investigation is to think alternately like a chemist and like a statistician, always beginning and ending like a chemist. The problem is a chemical one; the investigation is designed statistically; the experiments are carried out chemically; the results are analysed statistically; and then the chemist asks, do these results make sense – do they fit my understanding of chemistry? This book exists to introduce the statistical parts.

There are other techniques, specifically aimed at the *optimization* of a process or a product, which are essentially based on or derived from factorial experiments, but in which the total design is evolved in stages rather than being fully pre-planned. These are known by such names as *method of steepest ascent* and *evolutionary operations* (*EVOPS*). These follow a factorial or fractional factorial experiment to extend the investigation beyond the bounds initially set for the variables, and may then lead to a further factorial (or fraction) round some new suspected optimum set of conditions. EVOPS are particularly useful when the investigation has to be carried out on working chemical plant, where some variations can be permitted without measurable hazard to safety, efficiency, or other factor. It is possible to gain useful information in relatively few trials, but alternatively, if only small variations are possible, it may be necessary to run through the design a number of times before effects greater than random variation become apparent. There is no statistical analysis involved in these procedures.

In addition, a factorial-type experiment, conducted after a steepest ascent, may be carried out at several levels of the variables (a *composite design*). This leads, by least-squares fitting to polynomial equations, to the recognition of *response surfaces*.

These yield a more complete answer to questions such as:

How is the experimental result affected by the given variables over a range of settings of these variables?

What settings give a maximum (or minimum) result, and how rapidly does the result vary with modification of the settings?

What settings give a product satisfying the specification, which may have several requirements?

There are also a number of other designs which are popular because they are very effective for specific types of problem. Some of these are looked at in Sections 2.8 and 10.3. They are essentially fractional factorials in which it is assumed, rightly or wrongly, that interactions are negligible.

1.2 APPROACHES TO RESEARCH DESIGN

In essence, there are two possibilities:
(i) change one variable at a time – the classical method;
(ii) change one or more variables from one test to the next – statistical design.

Method (i) is necessary if one seeks to understand the fundamental relationship between cause and effect, and is the basis on which theories of equilibria and kinetics have been founded. Its importance in the development of chemical science is therefore undoubted. By understanding the relationship, one can endeavour to predict causes which, properly controlled, will produce wanted effects. Unfortunately, as will be shown in Sections 1.3 and 1.5, the information obtained is incomplete, and if interactions occur between the system variables, the predictions will not be reliable. Consequently, process modification may be difficult, or even bewildering. The method usually involves very detailed studies based on small quantities of carefully selected materials in the laboratory.

Method (ii) is the empirical one of directly measuring the effects of various changes in conditions without necessarily concerning oneself with the precise nature of the mechanisms producing the observed effects. It is therefore highly appropriate for process definition, process development, and optimization studies. This is particularly the case because one can carry out experiments on the actual industrial process or material being studied. On the other hand, the methods are applicable to any scale of work, and are equally at home in research, development, and analytical laboratories.

Statistical designs developed specifically for optimization purposes are not so structured as to give insight into what is happening in the chemical process: they are aimed simply at achieving a desired result with maximum economy. They only yield experimental results (they are said to provide *empirical* feedback) and the results determine the nature of subsequent experiments without consideration of chemical theory. Factorial and fractional factorial experiments, while not in principle as economical, are planned logically in advance and carried out in full. As a consequence of the analysis of variance (or other interpretation) which follows, they can yield *scientific* feedback, giving theoretical insights and therefore greater confidence to the experimenter. An example in Section 5.4 illustrates how it is possible intuitively to correlate obtained results with one's understanding of reaction mechanisms. It is important to try to interpret results in chemical terms, for statistical analysis necessarily implies some possibility (however small) that the results are in fact not due to the chemical factors the analysis has identified.

1.3 INVESTIGATING THE EFFECTS OF THREE CHEMICAL VARIABLES: THE ADVANTAGES OF STATISTICAL DESIGN

Suppose we wish to study the effect on (say) reaction yield of three variables A, B, and C (which can be anything from the familiar temperature, time and, concentration to, for example, reactor design, presence or absence of catalyst, flow rate, operators, order of addition of reagents, origin of raw material, and so on) and that we do this by operating the reaction at two levels of each variable (commonly called high and low).

If we use the *classical* approach we might

(a) run four trials at high A, keeping B and C low;
(b) run four trials at low A, keeping B and C low.

These eight experiments would give us four estimates of the effect (eff_A) of changing A from its low to its high value. But note that this would only be at low B and low C.

If we get a better yield at low A, we would standardize on that, and do

(c) four runs at high B, with low A and low C;
(d) four runs at low B, with low A and low C.

These eight experiments would give us four estimates of eff_B (but only at low A and low C).

If it has proved better to use high B, we would then do

(e) four runs at high C, with low A and high B;
(f) four runs at low C, with low A and high B.

These eight experiments give four estimates of eff_C (at low A and high B).

If we decide to use high C, we optimize on low A, high B, high C after *twenty-four* experiments. But we have not examined (*e.g.*) eff_A except at low B and C. And think of the other omissions.

If we use the *factorial design* approach, we proceed as follows:

There are eight possible combinations of the three variables, each at two levels as before (see Table 1.1).

These eight individual experiments (which statisticians often call 'treatment combinations') are the total number of possible combinations of three variables each at two levels. Since $8=2^3$, this is called a *2^3 factorial experiment*. Similarly, an investigation of four variables, each at three levels, leads to a 3^4 experiment with 81 treatment combinations (TCs).

One version of statisticians' symbols for the TCs (the version I personally find most useful) is shown in the right-hand column, and arises from the following convention:

> let the low level of any variable be denoted by (1); let the high level of any variable be denoted by its lower-case letter.

Also, multiply these symbols as in algebra.

Table 1.1 *Possible combinations of three variables at two levels*

Experiment	*Variable A*	*Variable B*	*Variable C*	*Treatment combination*
1	low	low	low	(1)
2	high	low	low	*a*
3	low	high	low	*b*
4	high	high	low	*ab*
5	low	low	high	*c*
6	high	low	high	*ac*
7	low	high	high	*bc*
8	high	high	high	*abc*

Hence, the experiment high A, low B, low C is called the treatment combination $a.(1).(1).=a$; while, *e.g.*, high A, high B, low $C=a.b.(1)=ab$.

These symbols are also used to represent the results of the individual TCs (the statistician calls them 'responses'). When you have some results to analyse, proceed by the Yates table, part of which is represented in Table 1.2 on p. 6.

1.3.1 Analysing the Results of Factorial Experiments

The Yates table method[2] calculates the mean estimates of every effect in a factorial design by a simple easy routine. All you have to do is:

(i) put down the results in a vertical list in the order shown in the right-hand column of Table 1.1 (no other order will do);

(ii) add them and subtract them in pairs for as many 'columns of analysis' as you have variables (here, three). That is to say, add the first two results, and put the answer at the top of the first analysis column. Do similarly with the second, third, and fourth pairs, to produce figures in the second, third, and fourth lines of the first analysis column. Then subtract the results in pairs (always first from second, third from fourth and so on) to put figures into the fifth to eighth lines of the first analysis column. Repeat this procedure twice.

Using the TCs symbols as results, your table would look like Table 1.2. The only bit you may not understand yet is the extreme right of column $\underline{3}$, but this is where it starts to get useful! To see why, read on.

1.3.2 Estimates of the Effects of Changing the Levels of Variables

Let us consider the estimates we have obtained of the effect on the result of changing A from low to high level. The difference in responses of any two experiments, one of which has 'a' in its TC, and the other of which has not (with no other difference between the TCs), is an estimate of the effect of A on the result (eff_A).

The estimates of eff_A are: $a-(1)$; $ab-b$; $ac-c$; $abc-bc$. So we have four estimates of eff_A from eight experiments.

Table 1.2 *Yates table analysis of a 2^3 factorial experiment (completed in Table 5.1)*

Response	Columns of analysis 1	2	3
(1)	$(1)+a$	$(1)+a+b+ab$	$(1)+a+b+ab+c+ac+bc+abc=$ Total
a	$b+ab$	$c+ac+bc+abc$	$a-(1)+ab-b+ac-c+abc-bc=4eff_A$
b	$c+ac$	$a-(1)+ab-b$	$b+ab-(1)-a+bc+abc-c-ac=4eff_B$
ab	$bc+abc$	$ac-c+abc-bc$	$ab-b-a+(1)+abc-bc-ac+c=4eff_{AB}$
c	$a-(1)$	$b+ab-(1)-a$	$c+ac+bc+abc-(1)-a-b-ab=4eff_C$
ac	$ab-b$	$bc+abc-c-ac$	$ac-c+abc-bc-a+(1)-ab+b=4eff_{AC}$
bc	$ac-c$	$ab-b-a+(1)$	$bc-abc-c-ac-b-ab+(1)+a=4eff_{BC}$
abc	$abc-bc$	$abc-bc-ac+c$	$abc-bc-ac+c-ab+b+a-(1)=4eff_{ABC}$

The mean $eff_A=\frac{1}{4}(a-(1)+ab-b+ac-c+abc-bc)$: compare the third column of analysis opposite the TC a in Table 1.2.

Similarly, we have four estimates of eff_B: $b-(1)$; $ab-a$; $bc-c$; $abc-ac$ [so mean $eff_B=\frac{1}{4}(b-(1)+ab-a+bc-c+abc-ac)$] and four estimates of eff_C: $c-(1)$; $ac-a$; $bc-b$; $abc-ab$, all from the same eight experiments.

Therefore we have as many estimates of eff_A, eff_B, and eff_C from *eight* statistically designed TCs as we obtained from *twenty-four* classical experiments! A factorial experiment is three times as efficient as a classical experiment on this count alone.

1.3.3 Bonus Information from Statistical Design: Estimating the Effects of Interactions

We have not by any means finished extracting information from our factorial experiments. To go further, we have to understand what the statistician means by *interactions*. You, as chemists, will find this easy. For instance, if you change (say) the time of refluxing of a chemical mixture from two hours to four hours, you would expect some change in yield. This is simply eff_{time}. Now, because reactions occur at different rates at different temperatures, you would be surprised if the same time change gave the same change in result at two different temperatures. The 'response curves' might look something like Figure 1.1, for an initial production of wanted product followed (alas!) by decomposition.

The effect of time change ($t_1 \rightarrow t_2$) is quite different at T_1 from what it is at T_2: we say that there is *time–temperature interaction*, and if $t=A$ and $T=B$, we write eff_{AB} as its symbol.

The interaction eff_{AB} may be defined as the effect of the level of B on eff_A, or as the effect of the level of A on eff_B, averaged over low C and high C. There are *four* estimates:

Effect of level of B on eff_A:

$(ab-b)-(a-(1))$ low C
$(abc-bc)-(ac-c)$ high C

Effect of level of A on eff_B:

$(ab-a)-(b-(1))$ low C
$(abc-ac)-(bc-c)$ high C

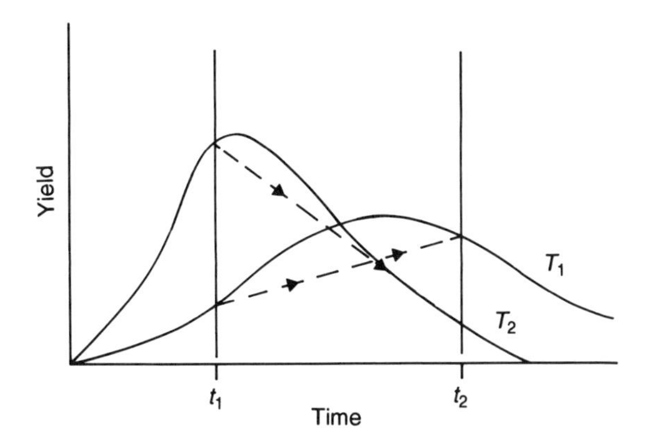

Figure 1.1 *Response curves showing two-factor interaction*

Each of these pairs sums to

$$ab - b - a + (1) + abc - bc - ac + c$$

and, as there are four estimates, we write

$$eff_{AB} = \tfrac{1}{4}(ab - b - a + (1) + abc - bc - ac + c)$$

in accordance with Table 1.2.

Since the classical research approach does not examine the effect of changing the level of any variable except at a fixed level of each other variable, it is impossible to detect interactions by it, unless many more experiments are done than were envisaged in Section 1.3. The classical method will not discover interactions if it is being used in the optimization mode described in that section.

So our simple 2^3 experiment gives us estimates of eff_A, eff_B, and eff_C, which the statistician calls the 'main effects' or the 'effects of the factors', and estimates of the interaction effects as well; and all for much less work. Factorial experiments are both more efficient and more informative than classical experiments are.

Example 1.1

Use these percentage yields from a given synthesis at two levels of variables A and B to calculate eff_A, eff_B, and eff_{AB}. Assume $A_1 < A_2$ and $B_1 < B_2$.

	A_1	A_2
B_1	42	47
B_2	57	62

Answer

The Yates analysis is:

		Columns of analysis	
TC	*Response*	1	2
(1)	42	89	208 = total response
a	47	119	$10 = 2eff_A$

b	57	5	$30=2eff_B$
ab	62	5	$0=2eff_{AB}$

(Note that the number of estimates of any effect is always equal to half the total number of experiments.)

So

$$eff_A=+5,\ eff_B=+15,\ eff_{AB}=0$$

Increasing A and B both increase the yield, with B more influential. There is no evidence of interaction.

Discussion

To increase the yield, it would be important to increase B more than A – see method of steepest ascent, Section 9.1. Lack of interaction means that the process will be easy to control in the experimental region investigated.

Example 1.2

Now consider the parts per thousand of a dangerous impurity in a product from a synthesis, at two levels of temperature T and time t. Assume $T=A$, $t=B$, and level 1<level 2.

	T_1	T_2
t_1	50	63
t_2	62	31

Answer

		Columns of analysis	
TC	*Response*	1	2
(1)	50	113	206
a	63	93	−18
b	62	13	−20
ab	31	−31	−44

$$eff_T=-9,\ eff_t=-10,\ eff_{AB}=-22$$

Discussion

Increase both T and t to decrease impurity. *The large interaction effect means* that the effect of either factor is very sensitive to the level of the other: *the process will be difficult to control*. On the other hand, the presence of *an interaction effect indicates that a maximum or minimum is somewhere about.*

Such maxima and minima (a minimum in Exercise 1.2) are pursued by optimization methods (see Chapter 9).

Having obtained the mean effects of the factors and interactions, the next step in many cases is to assess which, if any, of these effects are significant. Numerous significance tests are available, and some of these are considered in Chapter 2. After the Yates table, the effects are often subjected to an analysis of variance. The subject of factorial experiments, including a detailed example of the interpretation of results from such designs, is returned to in Chapter 5.

REFERENCES

1. W.E. Duckworth, 'Statistical Techniques in Technological Research – an Aid to Research Productivity', Methuen, London, 1968, p. 60.
2. F. Yates, 'Design and Analysis of Factorial Experiments', Imperial Bureau of Soil Science, London, 1937.

CHAPTER 2

Essentials of Statistical Analysis

2.1 STATISTICS – A VERY USEFUL TOOL FOR CHEMISTS

Typical situations in which chemists use statistical mathematics include:

(i) deciding whether there is a significant difference between the results of two or more processes,

(ii) deciding whether the effect of one variable depends significantly upon one or more others,

(iii) deciding whether some hypothesis is adequately confirmed by the experimental results.

Statistical methods have three essential purposes:

(a) the presentation of results in the clearest, simplest way (descriptive statistics and data presentation),

(b) the extraction of the maximum information from a given set of experiments,

(c) the derivation of correct conclusions in spite of variability in the experimental results.

Purpose (a) will not be considered in this book; readers who wish to pursue it will get help from books such as A.S.C. Ehrenberg's text 'A Primer in Data Reduction',[1] particularly its Part Five. With regard to purpose (b), we have already made a start in Chapter 1 on seeing how statistical principles may be used to design economical ways of carrying out experiments to obtain desired information. To facilitate purpose (c), we need essentially four things:

(i) a collection of relevant experimental results,

(ii) the mean value(s) of the results,

(iii) a quantitative representation of the variability of the results (most usefully, the standard deviation or its square, the variance),

(iv) techniques to assess the validity of our conclusions.

Items (ii) and (iii) give us estimates of the numerical characteristics of the *population*, the total material or system being studied, from the results obtained on the samples taken from it. We can then indicate the precision with which we can state these characteristics by quoting *confidence intervals* (*confidence limits*) within which the true value probably or very probably lies.

We usually wish to compare our conclusions with those which would arise from some hypothesis we have in mind, a hypothesis based on theory, previous work, or simple intuition. But if our sample results differ from the hypothesis, is the hypothesis wrong? Or are the sample results different only because of sampling error – *i.e.* our samples were not typical of the bulk? Provided we take samples *randomly* from the bulk, and do experiments in *random* order, we can give an answer to this problem, with some degree of confidence we can state, by using a *test of significance*. These tests are the techniques of item (iv) above. Such a test gives the probability that the difference between a sample value and a hypothesized value is due only to sampling error. If the probability is high, we accept the hypothesized value. If it is low, the difference is unlikely to have occurred by chance, *i.e.* it is probably a real difference, and we say the sample value is *statistically significant*. We then reject our hypothesis.

This hypothesis is called the *null hypothesis*, H_0. If we reject H_0, we accept the *alternative hypothesis*, H_1. For instance, if we are trying to improve the yield of a chemical process, the null hypothesis is that there is no improvement. The new mean yield μ equals the old one (μ_0), H_0: $\mu = \mu_0$. The alternative that we would be interested in is H_1: $\mu > \mu_0$. Note that there are other alternatives, H_1: $\mu < \mu_0$ and H_1: $\mu \neq \mu_0$, neither of which would interest us in this case. It is important to choose the right alternative hypothesis. We shall see in Section 2.6 that choosing the third alternative, $\mu \neq \mu_0$, doubles our uncertainty on given data, compared with either of the others. But sometimes we have to take the third alternative, *i.e.* when we are interested in a change of result in either direction (a *two-tailed test*). The other alternatives are *one-tailed tests*.

As with any other physical or mental tool we use in our work, we have to consider what statistics can and cannot do for us: its scope and limitations. Statistics *cannot* correct for inaccurate experimental results, and it *cannot* reveal conclusions not already implicit in the data. It *can* enable conclusions to be drawn which are not evident from simple observation of the data, and it *can* prevent false conclusions. It *cannot* lead to absolute certainty, but it *can* lead to conclusions drawn with different numerical degrees of certainty (confidence).

2.2 NECESSARY MEASUREMENTS, TYPES OF EXPERIMENTAL ERROR, ACCURACY, AND PRECISION

2.2.1 Average and Scatter of Results

Suppose we measure the length of a rod twice, and obtain the values 17.53, 17.51 cm. We usually take the average (mean) of these as our reported figure. The justification for this is that, when many repeat measurements are made, values nearer the middle of the *range* – *i.e.* the difference between the highest and lowest observed values – are more commonly observed (see Section 2.4 below) and we are more influenced by what we observe more commonly. But we have to ask whether mean values are sufficient evidence.

Suppose we now take two rods, and measure each five times, with the following results:

	Observed values (x/cm)	*Mean* (x/cm)
rod A	10.62, 10.59, 10.61, 10.61, 10.57	10.60
rod B	10.28, 10.31, 10.29, 10.32, 10.30	10.30

Because the *difference between the means*, $x_A - x_B$, is much greater than the variation in results on rod A or rod B (the scatter or *dispersion* of individual values), this looks like a real difference of rod lengths.

But if the results (giving the same means) are:

rod A	10.0, 10.7, 11.3, 9.8, 11.2	10.6
rod B	10.5, 9.6, 11.0, 10.6, 9.8	10.3

the difference between the means is less than the difference between individual values on a given rod. (This latter, by the way, is represented by a value known as the *difference within the means*, and we shall see that 'between' and 'within' comparisons are often important.) The variation between individual results leads to high uncertainty as to the truth of the matter.

It follows that

(i) comparison of means is not always sufficient evidence,

(ii) we also need to measure the scatter of results about their means,

(iii) we need a test or tests to help us decide whether the uncertainty caused by the variations between individual results is small – *i.e.* whether the difference between the means is significantly greater than the scatter among the results.

The same idea applies when we are comparing the mean and scatter of a set of experimental results with some fixed figure (say, a specification melting point or limit of impurity).

2.2.2 Types of Experimental Error, Accuracy, and Precision

The errors fall into two types:

(a) *Random errors* are due to failure to replicate a given procedure, leading to fluctuating results (*dispersion* of the data). The method, at least in the hands of that operator, is said to lack *precision*.

They are minimized, and therefore precision is maximized, by:

(i) finding a good technique,

(ii) practising it until proficient,

(iii) sticking to it like glue!

In statistical analysis, random errors are estimated by (represented by) *error variances* (sometimes known as *residual* variances, as they actually represent all unidentified sources of variation, not just procedural inconsistency).

(b) *Systematic errors* are due to a defect in materials, equipment, or method – including the operator – which occurs every time a given procedure is carried out. They lead to a mean different from the 'true' value, whatever that may be. The method is said to lack *accuracy*, or to have measurable *bias*.

Statistical analysis cannot detect (and therefore cannot estimate the significance of) difference of observed mean from true mean, for the latter can never be known. What it can detect and estimate is difference between experimental means, or between experimental mean and some fixed value.

Random errors are relatively easy to detect because of the varying results they produce. Systematic errors are more sinister, because they are experimentally more difficult to detect; they may indeed go undetected for long periods. They are minimized, and accuracy thereby maximized, by careful calibration of equipment and methods. As always, this is the comparison method of science: one's balance weights against standard weights, one's method against a standard method, one's method against standard samples, one operator or laboratory against another . . ..

2.3 INTRODUCTION TO TESTS OF SIGNIFICANCE

We saw in Chapter 1 that the Yates table gives us an easy, systematic way of calculating the magnitudes of the effects of changes in factor levels, and the associated interactions. Other manipulations of the experimental data would yield the same results. What we do not yet know is which of the factors and interactions, if any, have significantly large effects compared with the inevitable experimental error. Making this decision depends upon carrying out an appropriate significance test.

If only *two* samples or treatments are compared, the most common procedure is the *t-test*, which divides the difference between the means of two sets of numbers by what is known as the standard error of this difference. If the resultant figure is sufficiently large, it is likely that the two means (and therefore the two samples or treatments) are significantly different. The means may arise from two sets of experiments done in the same investigation, or from two laboratories; or one may be the accepted mean from many similar previous experiments, or may be the mean from an existing standard method with which a new procedure is being compared. If a significant difference is found in the last-named case, the results of the new method have a significant systematic error (a significant bias) compared with those from the standard. Tests such as the t-test are therefore tests for systematic errors, and, because they measure the relative positions of means, they are said to be measures of *location* of the results.

If *more than two* samples or treatments are to be compared, use of the t-test is cumbersome, since the calculation has to be repeated for each pair in turn. The hard labour can be (and is) taken out of this by computer programs, but this does *not* remove all objection, as there are cases where testing successive pairs of means does not produce clear evidence of significant differences.

The best procedure to compare more than two means is to use the *variance-ratio test* (the F-test). Since even the simplest factorial experiment, the 2^2, analyses for three effects, it follows that the F-test is the one to use in analysing factorials. We do in fact often want to compare the means of numerous samples, whether the design is a factorial one or not. For example, we might wish to examine the effects of three medical treatments on six groups of hospital patients: nine samples. We would measure some response – *e.g.* the proportion of patients discharged from hospitals in less than standard time – to decide:

(a) whether the treatment results really differ (do the treatment means differ significantly?)
(b) whether the groups of patients differ significantly across all the treatments (do the group means differ significantly?)
(c) whether there are significant treatment–group interactions (do the differences between treatment means differ significantly between groups of patients?)

Subdividing part of the total variation in the results into these separate categories, and assigning the rest as experimental error, is done by an *analysis of variance* (*ANOVA*). The individual effects are then compared with the experimental error by the F-test. The F-test is the principal test used in this book, because of its use in the analysis of factorial and related experiments. But the t-test is widely used, in both significance testing and the setting of confidence limits. Its uses are reviewed later in this chapter. We shall also see that graphical methods are very useful in identifying important effects. Each of these tests is related directly or indirectly to what is known as the Normal distribution of experimental data.

2.4 THE NORMAL (GAUSSIAN) PROBABILITY DISTRIBUTION

In many real-life cases of data collection, the property which is measured (the *variate*) will have only a finite set of possible values. Wage rates in a factory or shirt collar sizes in a department store are examples. Drawing a graph of such data would result in a stepwise histogram: a discontinuous distribution. In many other cases, particularly in scientific work, a continuous range of variate values can be envisaged. That is to say, the difference between two successive possible values is so small that it can be regarded as infinitesimal. The smallest difference between observable values depends only on the fineness (precision) of the observation procedure. The whole set of possible values in any distribution is said to be a *population*. The frequencies with which each value would be observed in the latter case, in an indefinitely long investigation, form a *continuous probability distribution*. Its graph would be some form of smooth curve.

One theoretical continuous distribution, the *Normal distribution*, is in many respects the cornerstone of modern statistical theory. It originated in the eighteenth century, when scientists observed an astonishing degree of regularity or pattern in observed values from repeated observations of a given quantity, such as the position of a given star. The great mathematical physicist Carl Gauss suggested that each reported value should be regarded as a combination of the true value and a term ε to represent the total error of observation.

$$x_{\text{observed}} = x_{\text{true}} + \varepsilon \tag{2.1}$$

He considered that ε is in fact the sum of a large number of small errors, each of which might be positive or negative, occurring at random.

Since it will rarely be the case that a large number of all-positive (or all-negative) errors occur together in making a single observation, very large total errors will be rare. In most cases, both positive and negative errors will occur, in different proportions from test to test, with corresponding degrees of mutual cancellation. Relatively small total errors will therefore be common. Hence, observed values near the mean will be common; observed values far from it will be rare.

On this laws-of-chance basis, Gauss showed that the frequency of variate values, y, lying between x and $x + \mathrm{d}x$ is given by $p(x)\mathrm{d}x$, in which $p(x)$, the *probability density function*, is given by

$$p(x) = \frac{1}{\sigma\sqrt{2\pi}} \exp\left(-\frac{x^2}{2\sigma^2}\right) \tag{2.2}$$

where σ is the standard deviation and σ^2 is the variance of the population of y values, and the mean value of y is zero.

But, generally, normal distributions have a non-zero mean. We denote such a *population* mean by the Greek letter μ. The probability density function then becomes

$$p(x) = \frac{1}{\sigma\sqrt{2\pi}} \exp\left(-\frac{(x-\mu)^2}{2\sigma^2}\right) \tag{2.3}$$

If the values of $p(x)$ from equation (2.3) are plotted against values of y with mean μ, scaled so that the total area under the curve is unity (the usual convention denoting total probability as unity), the normal distribution curve is obtained (Figure 2.1).

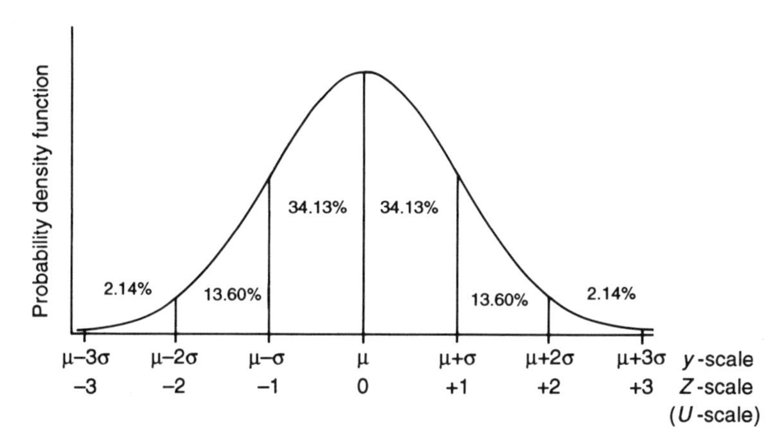

Figure 2.1 *The Normal distribution curve*

The lower scale on Figure 2.1 is obtained by deducting μ from every observed value, and dividing the result by σ. The result is said to be a *standard (or Z-) score*, and the distribution of those scores is the *standard Normal distribution*. Correspondingly, Z is sometimes called the *Normal variate, U*:

$$Z = U = \frac{x-\mu}{\sigma} \quad \text{for a single value} \tag{2.4a}$$

$$Z = U = \frac{\bar{x}-\mu}{\sigma_n} \quad \text{for a sample of } n \text{ tests} \tag{2.4b}$$

where $\bar{x}$ is the mean result and $\sigma_n = \sigma/\sqrt{n}$ (see Section 2.4.3). The purpose of using the standard distribution, with zero mean and unit standard deviation, is to ensure that only one table of U values is required. To use it, we must convert our observed scores to standard scores as indicated above. If we did not do so, we would face the impossible task of having a table for each possible combination of mean and standard deviation.

Figure 2.1 shows that if the standard deviation σ is large, the distribution curve must be wide. The observable values are scattered owing to large random errors. The curve has high *dispersion*, and the experiment has poor *precision*. Conversely, a highly precise experiment will give a sharply peaked curve (low dispersion, good precision). The *standard deviation of a population*, σ, is given by

$$\sigma = \sqrt{\frac{\sum_{i=1}^{N} (x_i - \mu)^2}{N}} \tag{2.5}$$

where there are N values, $x_1, x_2, \ldots x_i, \ldots x_N$, in the population. The square of this quantity is the population variance σ^2.

In the real world, of course, we do not sample every member of a population (every item in a batch). We usually have only a few values, n. This small sampling must add extra doubt to the uncertainty previously expressed by σ. The *population standard deviation calculated from a sample*, s, will therefore be larger than σ. It can be shown rigorously that s is given by

$$s = \sqrt{\frac{\sum_{i=1}^{n} (x_i - \bar{x})^2}{n-1}} \tag{2.6}$$

where $\bar{x}$ = sample mean $= x_1 + x_2 + \cdots\cdots x_n/n$.

s is commonly[2–4] called the *sample standard deviation* (although strictly that name should be applied to the quantity $\sqrt{[\Sigma(x_i - \bar{x})^2/n]})^5$, and correspondingly the square of this quantity, s^2, is commonly called the *sample variance*. For other than small values of n, calculation is simpler if equations (2.7) and (2.8) are used:

$$s = \sqrt{\frac{1}{n-1}\left\{\Sigma x^2 - \frac{1}{n}(\Sigma x)^2\right\}} \tag{2.7}$$

$$s^2 = \frac{1}{n-1}\left\{\Sigma x^2 - \frac{1}{n}(\Sigma x)^2\right\} \tag{2.8}$$

in which the term in parentheses is called the *corrected sum of squares*. The quantities s and s^2 are the *best estimates* of the population standard deviation and variance, and as such are given the symbols $\hat{\sigma}$ and $\hat{\sigma}^2$ in some texts. In the rest of this book, s and s^2 will be referred to simply as 'standard deviation' and 'variance', with the meanings given to them by equations (2.6), (2.7), and (2.8).

Example 2.1

Calculate the variance and standard deviation of a weighing procedure from the seven observations of the mass of an object:

12.6301, 12.6291, 12.6310, 12.6312, 12.6313, 12.6296, 12.6304 g.

Answer

(Make the calculation easier, and therefore less error-prone, by coding). Code: deduct 12.6000 from each value.

x	x^2
3.01×10^{-2}	9.0601×10^{-4}
2.91	8.4681
3.10	9.6100
3.12	9.7344
3.13	9.7969
2.96	8.7616
3.04	9.2416

$\Sigma x = 21.27 \times 10^{-2}$ $\Sigma x^2 = 64.6727 \times 10^{-4}$

$(\Sigma x)^2 = 452.4129 \times 10^{-4}$

$(\Sigma x)^2/7 = 64.6304 \times 10^{-4}$

$$s^2 = \tfrac{1}{6}\{64.6727 - 64.6304\} \times 10^{-4}$$
$$= 70.50 \times 10^{-8}\,\text{g}^2$$
$$s = 8.40 \times 10^{-4}\,\text{g}$$

N.B. The corrected sum of squares is a small difference between two large numbers. Hence there is need for careful calculation, and checking.

Exercise 2.1

Show that the estimated population standard deviation from ten batch yields, 89.6, 92.4, 84.7, 87.3, 94.3, 91.1, 88.3, 89.7, 89.8, 91.4 g, is 2.70 g.

2.4.1 Degrees of Freedom

The denominator $(n-1)$ in the expressions for s and s^2 represents the number of *degrees of freedom*, ϕ, of the data from which s and s^2 have been derived: *i.e.* the number of independent variables within a set of n variables which sum to a fixed value. For example, if we have to produce a hundred tons of a given product in five days, and we produce 18, 23, 19, and 17 tons on Monday to Thursday, we know we have to produce 23 tons on Friday. More generally, at least in theory, we can produce whatever we like on four of the five days, but we have no choice on the fifth day: our five daily results have four degrees of freedom. Similarly, a 2^3 factorial experiment has seven degrees of freedom.

Sixteen treatment combinations done as a single experiment have $\phi = 15$. But if they are divided into four blocks of four, then the results $x_1 \rightarrow x_{16}$ can be represented by Table 2.1.

Each block will have a standard deviation, as will the overall set of results. The degrees of freedom are divided as follows:

Table 2.1 *Division of an experiment into four blocks*

Block	1	2	3	4	
	x_1	x_5	x_9	x_{13}	
	x_2	x_6	x_{10}	x_{14}	
	x_3	x_7	x_{11}	x_{15}	
	x_4	x_8	x_{12}	x_{16}	*Grand Total*
Totals	$\Sigma x_{(1)}$	$\Sigma x_{(2)}$	$\Sigma x_{(3)}$	$\Sigma x_{(4)}$	$\Sigma x_{(1)}+\ldots.\Sigma x_{(4)}=X$

(i) Of the column totals, three can have any values; the fourth is fixed; hence there are three *between-blocks* degrees of freedom.

(ii) Any one of the column totals can be made up of three values of any size plus one fixed: thus within each block there are three *within-blocks* degrees of freedom.

The total degrees of freedom $=15=3$ between blocks $+$ $(4\times 3$ within blocks).

We have already seen the importance of comparing (difference between means) and (difference within means) in Section 2.2.1, and we shall repeatedly make this comparison in the analysis of variance. Correct partitioning of the degrees of freedom is therefore essential, since we use them to calculate the values of s and s^2. We also have to use correct degrees of freedom in other significance tests.

2.5 DEGREES OF CERTAINTY (CONFIDENCE) ABOUT THE MEAN OF EXPERIMENTAL RESULTS

2.5.1 Confidence Limits from a Single Observation

Referring again to Figure 2.1, the area under the curve between any two vertical bounds (limits) represents the percentage, or fraction, of all the possible values of y which fall between those bounds. Hence, for example, 68.3% of the values fall within one standard deviation on either side of the mean. More practically useful are statements such as: any one measurement of y, taken at random, is less than 5% likely to be more than $\pm 2\sigma$ from μ.

In this latter case we can say 'with 95.4% confidence' that μ is somewhere between $y=x-2\sigma$ and $y=x+2\sigma$. It is an unavoidable paradox that the greater degree of confidence we want to show, the wider these limits must be: we could say with 99.7% confidence that μ lies within $x\pm 3\sigma$. In this last case, we would expect that only three times in a thousand trials would μ actually lie outside those limits. Hence we have *confidence limits* (or *confidence intervals*) of population means as shown in Table 2.2.

So, if we obtain a single result x_1, we can be 80% sure that the population mean μ lies somewhere in the range $x_1\pm 1.28\sigma$, where σ would be the standard deviation from many measurements. Clearly, this can be useful only if the experiment which yields x_1 has followed many similar experiments.

But, usually, the chemist has not one but a number of replicate results from a

Table 2.2 *Confidence limits and percentage probabilities for a single observation*

Confidence limits $\mu \pm U\sigma$	*Percentage probability of obtaining a single result in this range*
$\mu \pm 1.28\sigma$	80.0
$\mu \pm 1.64\sigma$	90.0
$\mu \pm 1.96\sigma$	95.0
$\mu \pm 2.58\sigma$	99.0
$\mu \pm 3.29\sigma$	99.9

given investigation, giving a mean result $\bar{x}$. Also, σ may be considered known, but usually it is not. These two practical situations are considered in Sections 2.4.3 and 2.4.4.

2.5.2 Confidence Limits for μ from Sample Size n, with σ Known

It can be shown that if random samples, each of n units, are taken from a distribution with mean μ and standard deviation σ, the sample means will form a distribution having the same mean but with a smaller standard deviation $\sigma/\sqrt{n}$. Therefore, when we have n results, Table 2.2 applies, but with σ replaced by $\sigma_n = \sigma/\sqrt{n}$. The sample mean, $\bar{x}$, is our *best estimate* of μ, and the confidence limits are narrowed. σ_n is the *standard deviation of the mean* (otherwise known as the *standard error*; but see Section 2.5.3).

Example 2.2

The ten batch yields quoted in Exercise 2.1 were actually picked at random from many runs, and the whole investigation gave $\sigma = 2.30$. Show that the 80%, 90%, and 99% confidence limits of μ from the ten results are 88.97–90.83, 88.71–91.09, and 88.02–91.78 Respectively.

Answer

$\bar{x} = 89.9$ g, $\sigma_n = 2.30/\sqrt{10} = 0.73$

Percentage confidence	$\mu = \bar{x} \pm U\sigma_n$
80	$\bar{x} \pm 1.28\sigma_n = \bar{x} \pm 0.93$
90	$\bar{x} \pm 1.64\sigma_n = \bar{x} \pm 1.19$
99	$\bar{x} \pm 2.58\sigma_n = \bar{x} \pm 1.88$

Hence the confidence limits in the question are confirmed.

Discussion

Suppose a single batch yield was equal to the mean of the ten sampled: 89.9 g. The confidence limits from that single batch, with $\sigma = 2.30$, would be as in the table overleaf.

Percentage confidence	$\mu = x \pm U\sigma$
80	$x \pm 1.28\sigma = x \pm 2.94$
90	$x \pm 1.64\sigma = x \pm 3.78$
100	$x \pm 2.58\sigma = x \pm 5.93$

Comparing these two tables clearly demonstrates the narrowing of the confidence limits (the reduced uncertainty about the results) when an experiment is replicated. The necessary degree of replication in experimentation is discussed in Section 2.10.

Exercise 2.2

(a) Assume a single measurement on an electrochemical cell gave e.m.f. $= 0.721$ V; and assume a large number of corresponding measurements have given $\sigma = 0.02$ V. What is the upper limit of the 95% confidence interval? (**Answer** 0.725 V)

(b) But if eight measurements were made on the cell, giving as a mean the same 0.721 V, what then is the upper 95% limit? (**Answer** 0.722 V)

2.5.3 Confidence Limits from Sample Size n, with σ Unknown

In practice, we usually find that the population standard deviation is not known. All the information we have is from our sample of n tests or values. In such cases, we have to use the sample standard deviation s (with $n-1$ degrees of freedom) to calculate the standard error of the sample mean, $s_n = s/\sqrt{n}$. Whenever the two-word phrase, *standard error*, is used in the rest of this book, it is s_n which will be referred to.

We would then expect the 95% confidence intervals for μ to be of the form $\bar{x} \pm 1.96 s_n$, simply replacing σ_n by s_n in the formula (compare Section 2.5.2). For large samples ($n > 30$), we get very good approximate answers from this formula. But chemists very frequently have fewer – even far fewer – results than thirty.

As the number of results goes down, s becomes a poorer estimate of σ, and $\bar{x}$ is a poorer estimate of μ. Our uncertainty grows, and therefore our confidence limits must be wider. This is allowed for by using the more widely spaced *t-distribution* in place of the Normal distribution.

When the standard error s_n replaces σ_n in our calculations, we replace

$$U = \frac{|\bar{x} - \mu|}{\sigma_n} \tag{2.4}$$

by the *t-statistic*:

$$t_\alpha = \frac{|\bar{x} - \mu|}{s_n} = \frac{|\bar{x} - \mu|}{s/\sqrt{n}} \tag{2.9}$$

and the confidence limits for μ are then given by

$$\mu = \bar{x} \pm t_\alpha s_n \tag{2.10}$$

Table 2.3 *Some percentage points of the t-distribution for one-tailed tests*

	$\alpha=$ 0.10	0.05	0.025	0.01
$\phi=$ 1	3.078	6.314	12.706	31.821
4	1.533	2.132	2.776	3.747
7	1.415	1.895	2.365	2.998
10	1.372	1.812	2.228	2.764
20	1.325	1.725	2.086	2.528
40	1.303	1.684	2.021	2.423
∞	1.282	1.645	1.960	2.326

It can be shown that t-values have a distribution similar to, but more spread-out than the Normal distribution, thus suiting our increased uncertainty. Figure 2.2 compares the standard Normal distribution with the t-distribution for three degrees of freedom. The value of t_α depends upon the number of degrees of freedom, *e.g.* $\phi=(n-1)$ for a sample size n, and the required fractional confidence. Comparable to the standard Normal distribution, the t-distribution has a zero mean and unit standard error. Any t-value $+t_{\alpha,\varphi}$ has a fraction α of the total area under the curve lying above it; similarly there is a fraction α lying below $-t_{\alpha,\varphi}$. The values quoted in the t-table (an abstract from which is shown as Table 2.3) are derived from equation 2.9. The t-values are in rows corresponding to degrees of freedom, and in columns corresponding to α-values. A more complete table is shown in Appendix 3.

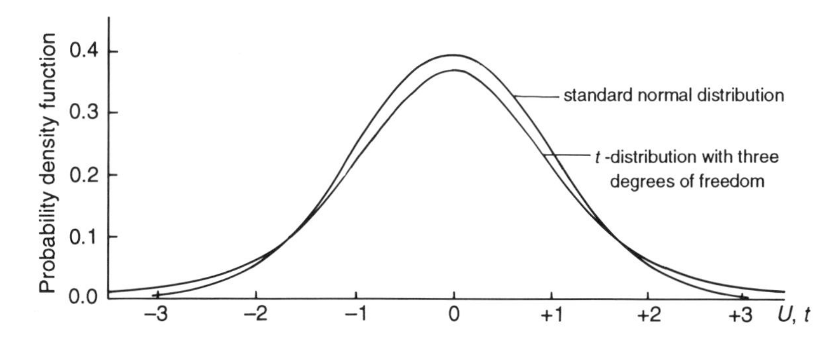

Figure 2.2 *Normal and t-distributions*

The table title customarily refers to the 'percentage points' of α, although the α-values in the table are always fractions of unity, as shown. *Note carefully that α refers to one tail of the distribution only*; the fractional chance of finding $\bar{x}$ outside any given confidence interval is twice α.

Example 2.3

A sample of eight observations, from a distribution of unknown μ, gave a sample mean $\bar{x}=10.40$ and a sample standard deviation $s=0.55$. Show that the 90%

confidence limits of μ are 10.04–10.76.

Answer

For $n=8$, $\phi=7$. From $s=0.55$, $s_n=0.55/\sqrt{8}=0.194$. The 90% confidence interval corresponds to $\alpha=0.05$. From the t-table, $t_{\alpha,\varphi}=t_{0.05,7}=1.895$.

The 90% confidence interval of $\mu=\bar{x}\pm t_{0.05,7}\cdot s_n$

$$=10.40\pm 1.895\ (0.194)$$

$$=10.04\text{–}10.76$$

Exercise 2.3

Calculate the 95, 99, 99.9% confidence limits for μ where $\bar{x}$, s, $n=$ (a) 15.0, 5.0, 10; (b) 25.0, 3.0, 23.

Answers

(a) 11.43, 18.57; 9.86, 20.14; 7.45, 22.55
(b) 23.70, 26.30; 23.24, 26.76; 22.63, 27.37

2.5.4 Confidence Limits for the Difference between Two Independent Means

Sometimes we wish to compare two sets of data which have been obtained separately from each other. One example would be samples from two laboratories or plants, working independently of one another. Another would be two sets of samples: one is tested immediately, but the other is subjected to some extra treatment before testing, and the two sets are then compared. A change in mean can be found, but changes in individual specimens are not known. The procedure for such *independent means* is as follows.

Consider two random samples, of n_1 and n_2 units and means $\bar{x}_1$ and $\bar{x}_2$. For this situation, we write:

$$t=\frac{|\bar{x}_1-\bar{x}_2|}{(\text{standard error of the difference between the means})} \qquad (2.11)$$

The *standard error of the difference between the means, SE*, is the square root of the sum of the squares of the standard errors $s_{n(1)}$ and $s_{n(2)}$. Hence,

$$t_\alpha=\frac{|\bar{x}_1-\bar{x}_2|}{\sqrt{s_{n(1)}^2+s_{n(2)}^2}} \qquad (2.12)$$

with total degrees of freedom $=n_1+n_2-2$.

If the two samples have the same variance, so that $s_1=s_2=s$, this transforms into

$$t=\frac{|\bar{x}_1-\bar{x}_2|}{s\sqrt{\dfrac{n_1+n_2}{n_1n_2}}} \qquad (2.13)$$

where

$$s=\sqrt{\frac{\Sigma(x_1-\bar{x}_1)^2+\Sigma(x_2-\bar{x}_2)^2}{n_1+n_2-2}} \tag{2.14}$$

$$=\{[(n_1-1)s_1^2+(n_2-1)s_2^2]/n_1+n_2-2\}^{\frac{1}{2}} \tag{2.14a}$$

The confidence limits are given by

$$\mu_1-\mu_2=\bar{x}_1-\bar{x}_2\pm t_\alpha s\sqrt{\frac{n_1+n_2}{n_1 n_2}} \tag{2.15}$$

When $n_1=n_2=n$, this becomes

$$\mu_1-\mu_2=\bar{x}_1-\bar{x}_2\pm t_\alpha s\sqrt{\frac{2}{n}} \tag{2.16}$$

Note: This bringing together of two variances into a single estimate (the *pooling of variances*) is valid only if s_1 and s_2 are the same, or nearly so. If this assumption cannot be made (the proper thing to do is to carry out the F-test of the variances to see if they are significantly different), equation (2.12) is used, but the value of ϕ in consulting the t-table is not (n_1+n_2-2) but comes from the formula

$$\frac{1}{\phi}=\frac{1}{\phi_1}\left\{\frac{s^2_{n(1)}}{s^2_{n(2)}+s^2_{n(2)}}\right\}^2+\frac{1}{\phi_2}\left\{\frac{s^2_{n(2)}}{s^2_{n(1)}+s^2_{n(2)}}\right\}^2$$

2.5.5 Confidence Limits for the Difference between Related Means

As an alternative to the second example in Section 2.5.4, where two sets of specimens were treated differently, we could test *one* given set of specimens before and after treatment. In this test–retest situation, we have a *test of related means*. Another example is when we have several batches of material, and we analyse each batch by two methods. We say we have subjected our experimental design to *blocking*; in these examples, each block consists of two tests. Here we have the difference within each pair of measurements on a given specimen, which we did not have before, and it is this difference which is of interest. Because the objects examined are the same in both runs, we can expect less random variation and correspondingly a more sensitive (a more precise) test for a change in mean. But this last effect is offset by a reduction in degrees of freedom; since n samples, each tested twice, yield n differences of results, ϕ is now not $(n+n-2)$ but $(n-1)$.

In this case, t_α is defined by:

$$t_\alpha=\frac{|\bar{x}_1-\bar{x}_2|}{s_d/\sqrt{n}}=\frac{\bar{d}}{s_d/\sqrt{n}} \tag{2.17}$$

where

$\bar{d}$ = mean difference within pairs
s_d = *standard deviation of the differences within pairs*

$$= \sqrt{\frac{\Sigma d^2 - (\Sigma d)^2/n}{n-1}}$$

in which d = difference within any pair.

The confidence limits are therefore

$$\mu_1 - \mu_2 = \bar{x}_1 - \bar{x}_2 \pm t_\alpha s_d/\sqrt{n} \tag{2.18}$$

These two cases are illustrated in Examples 2.4 and 2.5.

Example 2.4: independent means

Suppose ten specimens of an alloy are tested for tensile strength, and the mean value $\bar{x}_1$ found. Ten further specimens are subjected to a standard process of heat treatment, and the second mean $\bar{x}_2$ found. Let the results be as in Table 2.4.

Table 2.4 *Tensile strength of alloy samples*

Specimen	*Test 1; x_1*	*Test 2; x_2*	$x_2 - x_1 = d$
1	20.7	21.9	1.2
2	18.4	20.8	2.4
3	19.8	21.1	1.3
4	18.8	20.1	1.3
5	19.9	19.9	0
6	23.4	24.4	1.0
7	23.7	25.5	1.8
8	20.8	21.6	0.8
9	20.0	24.6	4.6
10	22.0	23.4	1.4
Totals	$\Sigma x_1 = 207.5$	$\Sigma x_2 = 223.3$	$\Sigma d = 15.8$
Means	$\bar{x}_1 = 20.75$	$\bar{x}_2 = 22.33$	$\bar{d} = 1.58 = \bar{x}_2 - \bar{x}_1$

For these two random samples, $n_1 = n_2 = 10$, $\phi = n_1 + n_2 - 2 = 18$.

If they both come from populations with the same variance, the overall standard deviation, s, is as in equations (2.14) and (2.14a), equivalent to (and most easily calculated as)

$$s = \left\{ \frac{1}{n_1 + n_2 - 2} \left[\Sigma x_1^2 - \frac{(\Sigma x_1)^2}{n_1} + \Sigma x_2^2 - \frac{(\Sigma x_2)^2}{n_2} \right] \right\}^{\frac{1}{2}} \tag{2.19}$$

and the standard error of the difference between means, SE, is

$$SE = s\sqrt{\frac{n_1 + n_2}{n_1 n_2}}; \text{ here} = s\sqrt{\frac{2}{n}} \tag{2.20}$$

From the twenty x values, $\Sigma x_1^2 = 4334.43$, $\Sigma x_2^2 = 5022.37$, $(\Sigma x_1)^2/10 = 4305.63$, $(\Sigma x_2)^2/10 = 4896.29$.

As a consequence, $s = 1.896$ and $SE = 0.848$.

From the t-table, for $\alpha=0.025$ and $\phi=18$, $t=2.10$. $\alpha=0.025$ corresponds to the *95% confidence limits*, which must be

$$\begin{aligned}\mu_2-\mu_1 &= \bar{x}_2-\bar{x}_1 \pm 2.10(0.848)\\ &= 1.58 \pm 1.78\\ &= -0.20 \rightarrow +3.36\end{aligned}$$

If a confidence interval is entirely positive or entirely negative, we can have confidence that $\mu_2-\mu_1$ is not zero. These results do not give us that confidence. The heat treatment has not been shown to have significant effect, even though the mean observed effect was quite measurable.
Note: another useful way of looking at confidence intervals – even if they include zero – is to ask if either limit makes a technologically or economically interesting difference between means. This topic is dealt with in Section 3.1.

Example 2.5: related means

Now assume that the results shown in Table 2.4 in Exercise 2.4 were obtained when the same ten marked specimens were tested before (x_1) and after heat treatment (x_2). In the experiment, a *paired comparison test*, we calculate $\Sigma d^2=38.58$, $(\Sigma d)^2/n=249.6/10=24.96$. Then the standard difference s_d is given by

$$s_d=\sqrt{\frac{\Sigma d^2-(\Sigma d)^2/n}{n-1}}=\sqrt{\frac{38.58-24.96}{9}}$$

$$=1.23$$

and

$$s_d/\sqrt{n}=0.389.$$

From the t-table, $t=2.26$ for $\alpha=0.025$, $\phi=9$. So the 95% confidence limits are now

$$\begin{aligned}\mu_2-\mu_1 &= \bar{x}_2-\bar{x}_1 \pm t_\alpha s_d/\sqrt{n}\\ &= 1.58 \pm 2.26\ (0.389)\\ &= 0.70 \rightarrow 2.46\end{aligned}$$

If we compare these confidence limits for the same data treated as independent means ($-0.20 \rightarrow +3.36$, Example 2.4) we see why testing for related means was described as a 'more precise' test in the first paragraph of this Section.

For the related means, the 99% limits are also both positive: 0.32–2.84. We can therefore conclude with high confidence that $\mu_2-\mu_1$ is not zero: there has been a significant increase in tensile strength. The formal way of stating this is to say that the *null hypothesis*, $\mu_1=\mu_2$, has been rejected, and the *alternative hypothesis*, $\mu_2>\mu_1$, has been accepted.

Discussion

We have reached different conclusions, even though we used the same figures. This indicates that *testing in (related) pairs* allows for more sensitive testing of an effect, since it reduces the random variation in the data. We shall come to this again when considering blocking in the design of experiments. Examples 2.4 and 2.5 constitute a simple example of how better design gives more meaningful results.

Exercise 2.4: independent means

Ten students are drawn at random from each of two large groups of students, A and B, and each carries out a given synthesis. Suppose their yields are as follows:

Group	*Yield/g*	*Mean yield/g*
A	47 42 40 39 33 28 25 22 20 14	31.0
B	47 46 45 42 37 36 31 25 25 16	35.0

(i) Show that the variances for the two groups are $s_A^2 = 118.0$, $s_B^2 = 110.7$.
(ii) By *F*-test, these variances are not significantly different, so it is valid to calculate confidence limits for the difference between the population mean yields, μ_1 and μ_2, from

$$|\mu_A - \mu_B| = |\bar{x}_A - \bar{x}_B| \pm t_\alpha s \sqrt{\frac{2}{n}}$$

where

$$n = n_1 = n_2 = 10$$

and

$$s = \left\{\frac{1}{2n-2}\left[\Sigma x_1^2 - \frac{(\Sigma x_1)^2}{n} + \Sigma x_2^2 - \frac{(\Sigma x_2)^2}{n}\right]\right\}^{\frac{1}{2}}$$

Calculate the 95% confidence limits (*i.e.* using $t_{0.025,18}$) using equation (2.16) to give an opinion on the likelihood that the two groups of students are significantly different (*i.e.* $|\bar{x}_A - \bar{x}_B|$ is unlikely to be zero).

Answer (part ii)

$$t_{0.025,18} = 2.10; \; s = 10.69$$

so

$$|\mu_A - \mu_B| = |\bar{x}_A - \bar{x}_B| \pm \sqrt{0.2}st_\alpha = -6.0 \rightarrow +14.0$$

No significant difference has been demonstrated between the two groups, since the confidence limits include zero.

The null hypothesis is accepted. The differences within groups are so big that the difference between groups is not significant.

Exercise 2.5: related means

Now suppose that the same two sets of results were obtained, but the group B students were the group A students after specialist tuition in the synthesis technique, and that the pairs of results were as follows:

Student	1	2	3	4	5	6	7	8	9	10
Run A	47	42	40	39	33	28	25	22	20	14
Run B	46	45	47	37	42	36	31	25	25	16

Again, $\bar{x}_A = 31.0$, $\bar{x}_B = 35.0$.

Use equation (2.18) for $\alpha = 0.025$, $n = 10$, to show whether the tuition probably had a significant effect.

Answer

$t_{0.025,9} = 2.26 \qquad s_d/\sqrt{n} = 1.16$

$\mu_B - \mu_A = 4.0 \pm 2.26(1.16) = 1.38 \rightarrow 6.62$

The 95% confidence interval is entirely positive. The null hypothesis $\mu_B - \mu_A = 0$ is rejected. It is accepted that the tuition has had a significant effect.

Discussion

Exercises 2.4 and 2.5 again demonstrate that related means (paired comparisons) give a more sensitive test of significance than do independent means: they narrow the confidence interval, and thus give a more precise idea of the situation.

2.5.6 Other Confidence Intervals

Variance (or its square root, standard deviation) is used to represent the scatter of a set of experimental results about their mean. Just as we wish to estimate the population mean from the experimental mean, so we may wish to estimate the population variance from the experimental variance. By so doing, we may forecast the variation to be found within a large batch of items by examining relatively few of them. High-quality production requires tight confidence limits in respect of both the mean and the variance. Tightening both sets of limits is the 'never-ending search for quality improvement' much spoken of in recent literature.

From the sample variance s^2, for a sample of n items, the $100(1-2\alpha)$ percent confidence interval for the population variance σ^2 is given by

$$\frac{(n-1)s^2}{\chi^2_{\alpha,n-1}} < \sigma^2 < \frac{(n-1)s^2}{\chi^2_{100-\alpha,n-1}} \tag{2.21}$$

So, for example, the 95% confidence limits are

$$\frac{(n-1)s^2}{\chi^2_{0.025,n-1}} < \sigma^2 < \frac{(n-1)s^2}{\chi^2_{0.975,n-1}}$$

and we would be wrong on five percent of occasions if we said that σ^2 lies within these limits. χ^2 (chi-squared, pronounced kigh-squared) is represented by

$$\chi^2 = (n-1)s^2/\sigma^2 \tag{2.22}$$

and values can be found in standard sets of statistical tables.

In a similar way, we may wish to indicate how confident we are about a numerical relationship (a regression) we have established between variables. For instance, if we have taken n pairs of observations $(x_1,y_1)\rightarrow(x_n,y_n)$, we may use the method of least squares to set up the equation $y=a+bx$, where a and b have such values that the variance of y values about the line is minimized. Since two constraints (a and b; or $\bar{x},\bar{y}$ with either a or b) determine the line, the equation has $(n-2)$ degrees of freedom for n experimental points. For a point i on this line, $y_i=a+bx_i$. But the corresponding experimental value of y may include an 'error' ε – *i.e.* a deviation from that line – so that experimentally $y_i=a+bx_i+\varepsilon_i$. As a consequence, the variance of $y, s^2_{y/x}$, is given by

$$s^2_{y/x}=\frac{\Sigma\varepsilon_i^2}{n-2}=\frac{\Sigma(y_i-a-bx_i)^2}{n-2} \tag{2.23}$$

It can be shown that the $100(1-2\alpha)$ confidence interval for y, related to any stated value of $x=x_0$, is given by

$$y=a+bx_0\pm t_{\alpha,n-2}s_{y/x}\sqrt{\frac{1}{n}+\frac{(x_0-\bar{x})^2}{\Sigma(x_i-\bar{x})^2}} \tag{2.24}$$

The confidence limits are two smooth curves, one on either side of the chosen regression line: see Figure 2.3. Note that the confidence interval, and therefore the uncertainty, is smallest when $x_0=\bar{x}$.

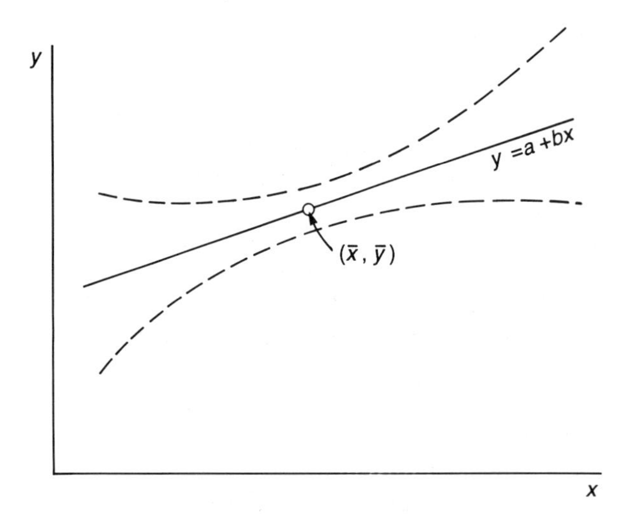

Figure 2.3 *Confidence intervals for $y=a+bx$*

These intervals are not pursued further in this book, but are discussed in detail in standard textbooks of statistical analysis.[2–4]

2.6 TESTS OF SIGNIFICANCE

When we calculated confidence limits, we took values from the t- or U-tables and inserted them in the appropriate equations. In tests of significance, we use table values in another way. In the *t-test of significance*, for example, we compare t-values, calculated from our own data using equations such as (2.9), (2.11), (2.12), (2.13), or (2.16), with values in the table for the appropriate number of degrees of freedom. This identifies the α-value for our experiment, which is the fractional chance that our results could have arisen merely by random error, without $\bar{x}$ and μ (or $\bar{x}_1$ and $\bar{x}_2$) truly being different. The α-values refer to one tail of the distribution only: see Figure 2.4.

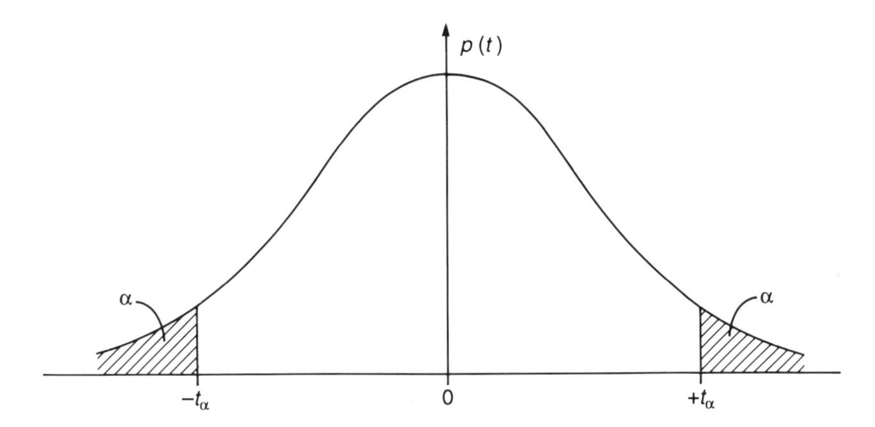

Figure 2.4 *The t-distribution*

If we are interested only if t_α is positive (or, equally, only if t_α is negative) and *we state this before we examine the data*, this is a *one-tailed* test, and the chance probability is α. If we do not make such a statement, this is a *two-tailed* test, and the probability of our results occurring by chance is 2α. Often, we can opt for a one-tailed test (as, for instance, when we are seeking to increase the yield or reaction rate of a process, or to decrease its cost), but sometimes we have to use a two-sided test (as, for instance, when we investigate whether a new analytical method gives a mean result different from that of an established method). In the two-tailed test, sample means which fall into either tail of the distribution lead to the null hypothesis being rejected. In a one-tailed test, only means falling in the nominated tail lead to rejection.

Choosing a one-tailed test, where legitimate, halves the chance probability and consequently increases the likelihood of our observing a significant result. Table 2.5 shows the experimental (t_e) values corresponding to various percentage probabilities of chance occurrence. From it we see, *e.g.*, that for 95% confidence with five degrees of freedom, we need $t_e = 2.02$ on a one-tailed test or $t_e = 2.57$ on a two-tailed test. Smaller differences between means give significance on one-tailed tests.

These so-called *critical values* of t are connected with the need to distinguish between highly significant, significant, or less significant results. How these are discriminated is somewhat a matter of choice, but commonly the distinctions are as shown in Table 2.6.

Hence, $\alpha = 0.04$ is significant on a one-tailed test, but only possibly significant on its two-tailed equivalent.

Table 2.5 *Some critical values of t*

	% Probability level		*Degrees of freedom ψ*				
α	*one-tailed test*	*two-tailed test*	*1*	*5*	*10*	*20*	*30*
0.10	10.0	20.0	3.08	1.48	1.37	1.33	1.31
0.05	5.0	10.0	6.31	2.02	1.81	1.73	1.70
0.025	2.5	5.0	12.7	2.57	2.23	2.09	2.04
0.005	0.5	1.0	63.7	4.03	3.17	2.85	2.75

The percentage probability levels are the percentage chances that the observed results could be obtained merely by random sampling.

Table 2.6 *Attribution of significance of p-values*

p%	*Significance of difference between means*
<1.0	Highly significant
1.0–5.0	Significant
5.1–10.0	Possibly significant (continue testing)
>10.0	Not significant

This last example may make it clear why emphasis was laid on declaring a one-tailed or a two-tailed test before data analysis. There is always a tendency to strive for significant results, and it is unacceptable to analyse the data first, and then declare the type of test to force up the level of significance. The type of test should be determined by the objective of the experimentation, and therefore the declaration can indeed be made before any work starts. If we declare a one-tailed test, but the results are in the other direction, our experimental work (or our theoretical understanding of the process) needs improvement, and statistical analysis is superfluous for the present. See Section 2.6.2 for further discussion.

2.6.1 Use of the Normal Distribution Table

If we already have so much information that we feel we know the population mean μ and the population standard deviation σ, we can proceed as in Examples 2.6 and 2.7. This is often the case in quality control work, where a large amount of earlier data is available.

Example 2.6: to test whether a single observation could come from a Normal population

Many batches of a product have been made in a chemical plant, and it is known from this long experience that the combined water in the product is distributed Normally with $\mu=2.48\%$ and $\sigma=0.12\%$. After necessary repairs to the plant, a single test on the next batch gives 2.76% moisture.

For this batch, $U_e = (x - \mu)/\sigma = 0.28/0.12 = 2.33$. The Normal distribution table (Appendix 2) shows the 100α percentage points of the standardized Normal variate U. Thus U_α is the value of the variate which has probability (100α)% of being exceeded merely by random sampling. For this example, and Example 2.7 which follows, the table shows

(100α)%	U
1.0	2.326
0.9	2.366

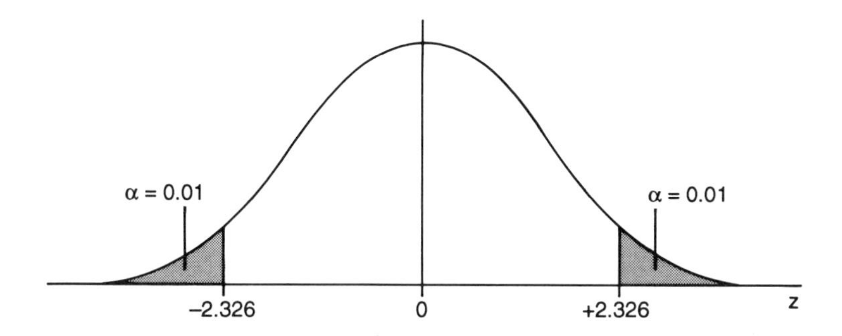

Figure 2.5 *Tails of the Normal distribution*

Therefore, there is only 2% chance of the moisture content being found as 2.76% without the repairs having caused some variation in the product. [Note that we have to use the two-tailed probability level, because we did not state in advance that only results higher (or lower) than before would be of interest].

Example 2.7: to test whether a sample mean $\bar{x}$ differs significantly from μ

Suppose that, instead of taking just one sample from the first batch (as in Example 2.6), it was decided to run four batches, take one sample from each, and regard the four samples as representative of the new product. And suppose the mean combined water from these was found to be 2.62%. Then,

$$U_e = \frac{2.62 - 2.48}{0.12/\sqrt{4}} = \frac{0.14}{0.06} = 2.33$$

There is again less than 2% chance of the mean moisture content being found as 2.62% over four batches, as mere random variation of the old process. Comparing Examples 2.6 and 2.7 again reminds us that if we take several samples, instead of just one, then smaller differences of results are needed to show a given level of significance. Our analyses are more sensitive of changes when we take several samples.

Exercise 2.6

The population mean strength of a standard alloy is $\mu = 49.3 \times 10^4$ kPa, with standard deviation $\sigma = 0.68 \times 10^4$ kPa. Suppose a change is made in the alloy composition, and five samples of the new product are tested, giving the following yield strengths:

$$49.9,\ 51.0,\ 49.7,\ 50.7,\ 49.0\ (\times 10^4\ \text{kPa})$$

Has the yield strength been significantly improved?

Answer

$$\bar{x} = 50.1 \times 10^4;\ U = 2.67$$

From the Normal distribution table, probability $p(U > 2.67) < 0.4\%$. The yield strength has been significantly improved.

Exercise 2.7

Find whether $\bar{x}$ and μ differ significantly, and state at what two-tailed probability level, when

$\mu, \sigma, \bar{x}, n =$ (a) 83.0, 5.7, 80.0, 7
(b) 46.0, 11.4, 35.2, 8

Now find one-tailed probabilities for

$\mu, \sigma, \bar{x}, n =$ (c) 62.1, 2.1, 63.3, 7
(d) 5.50, 0.11, 5.31, 4

Answers

(a) $0.20 > 2\alpha > 0.10$ Not significant
(b) $0.006 > 2\alpha > 0.004$ Highly significant
(c) $0.10 > \alpha > 0.05$ Possibly significant
(d) $0.001 > \alpha > 0.0001$ Highly significant

2.6.2 Significance in One-tailed Tests

Please note that the results in Exercise 2.7 (c) and (d) are significant *only if* the tail of interest in (c) is that $\bar{x}$ should be greater than μ, and in (d) that $\bar{x}$ should be less than μ. If the other tail had been of interest, the results would not have been significant, irrespective of however small α might be. In general, a one-sided test is appropriate if a deviation in the direction opposite to that tested would have no practical importance. For instance, if a specification imposes an impurity limit on a product, it would be important if it were significantly higher but not if it were significantly lower.

2.6.3 The *t*-Test

Use of the Normal test of significance applies only when σ is known exactly, or at least is based on a sample size $n \nless 30$. When this is not the case, we must use the sample standard deviation s in place of σ, and U is replaced by t.

Example 2.8: to test whether a sample mean $\bar{x}$ differs from an expected value E when σ is estimated from the sample

From our experimental data $x_1, x_2, \ldots x_n$, we calculate $\bar{x}$ and s, and from them obtain our experimental t-values, t_e:

$$t_e = \frac{|\bar{x} - E|}{s/\sqrt{n}} \tag{2.9}$$

We compare this value, with $\phi = n-1$, with table values (see Tables 2.3, 2.5 and Appendix 3) to obtain the fractional probability, α or 2α, that the results could have occurred by chance. We can then state whether the results are significant (*i.e.* whether the null hypothesis is rejected), and if so, with what percentage confidence, $100(1-\alpha)$ or $100(1-2\alpha)$.

Specimen calculation

The structure of a chemical compound is such that it contains 7.15% X when pure. Ten analyses by a standard method of negligible systematic error gave the following data:

$\bar{x}$	E	s	n
6.94	7.15	0.31	10

Is $\bar{x}$ significantly different from E?

Answer

$$t_e = \frac{0.21}{0.31/\sqrt{10}} = 2.14$$

As we have *not* asked whether $\bar{x}$ is significantly *smaller*, we must use a two-tailed test. For this, with $\phi = 9$, $t = 2.14$ corresponds to $0.10 > 2\alpha > 0.05$. We can say, with better than 90% confidence but less than 95% confidence, that $\bar{x}$ is different from E. This result is 'possibly significant'. If there had been an *a priori* statement expecting $\bar{x} < E$, we could have applied a one-tailed test, giving $0.05 > \alpha > 0.025$. This is a 'significant' result; we could say with better than 95% confidence that $\bar{x}$ is smaller than E.

Exercise 2.8

A chemical compound contains 3.40% Y when pure. Ten analyses by an unbiased method gave

3.67, 3.20, 3.42, 3.45, 3.14, 3.14, 3.21, 3.05, 3.11, 3.21% Y.

Is the percentage of Y in the product significantly low?

Answer

Here the question allows a one-tailed answer.

$t_e = 2.30$. For $\phi = 9$, $\alpha < 0.025$.

With greater than 97.5% confidence, the product is deficient in Y.

Example 2.9: to test the difference between the means of two samples from the same population

If two samples, of n_1 and n_2 items, are taken at random from a single population, we can calculate t_e by using equation (2.13):

$$t_e = \frac{|\bar{x}_1 - \bar{x}_2|}{s\sqrt{\dfrac{n_1+n_2}{n_1 n_2}}} \tag{2.13}$$

where

$$s = \left\{ \frac{1}{n_1+n_2-2} \left[\Sigma x_1^2 - \frac{(\Sigma x_1)^2}{n_1} + \Sigma x_2^2 - \frac{(\Sigma x_2)^2}{n_2} \right] \right\}^{\frac{1}{2}}$$

We compare this value, with $\phi = (n_1 + n_2 - 2)$, with table values to obtain the fractional probability that the results could have occurred by chance.

Specimen calculation

Suppose we have the following results:

$\bar{x}_1$	$\bar{x}_2$	n_1	n_2	s
110	126	6	8	11.8

Then

$$t_e = \frac{16}{11.8\sqrt{14/48}} = 2.51$$

From the t-tables, for $\phi = 12$,

Probability	*two-tailed*	10	5	1	0.1
level, %	*one-tailed*	5	2.5	0.5	0.05
t		1.78	2.18	3.05	4.32

If our alternative hypothesis had been $\bar{x}_1 \neq \bar{x}_2$, we would use a two-tailed test, where $t_e = 2.51$ corresponds to $5\% > 2\alpha > 2.5\%$. The two means are significantly different.

If we had stated our alternative hypothesis as $\bar{x}_1 < \bar{x}_2$, the difference between means is significant at better than (less than) the 2.5% confidence level.

Exercise 2.9

Use equations (2.13) and (2.14) to test the following data for significance (i) two-tailed and (ii) one-tailed for $\bar{x}_2 > \bar{x}_1$:

	$\bar{x}_1$	$\bar{x}_2$	n_1	n_2	s	s_1	s_2
(a)	47.3	56.6	16	9	9.9	—	—
(b)	102.0	110.7	9	11	—	9.3	9.8

Answers

(a) $t_e = 2.25$ (i) $0.05 > 2\alpha > 0.02$; significant at better than 5% level
(ii) $0.025 > \alpha > 0.01$; significant at better than 2.5% level
(b) $t_e = 1.69$ (i) $0.20 > 2\alpha > 0.10$; not significant
(ii) $0.10 > \alpha > 0.05$; possibly significant

Example 2.10: *t*-test for paired comparisons

When comparing two different methods, experiments are often carried out in pairs. For example, we may wish to compare two analytical methods for general applicability to determining a particular element in its compounds. The variation of percentage content between compounds will be much bigger than any likely difference between the results of the two methods. We therefore apply both methods to each compound, and compare the two results on any one compound. Another example is where the performances of two production plants are being compared. If several trials are carried out in each plant, it may not be possible to provide sufficient homogeneous raw material to cover the whole investigation. However, provided a single batch of raw material is sufficient for one trial in each plant, each pair of such trials will give an estimate of the difference between plants. We can compare their performances even though the batch-to-batch variation (difference within plants) may be greater than the between-plants effect.

In general, we use the paired comparisons method if we suspect that the within-means variation may be big enough to interfere with our assessment of the between-means difference. In such cases, as in Section 2.5.2, the quantity examined is the difference within the pairs of results. We compare the mean difference between the two methods, or plants, with its standard deviation $s_d/\sqrt{n}$, where s_d is the standard deviation of the observed differences in n pairs (equation 2.17).

Specimen calculation

Eight compounds were analysed in order to compare two methods for determining percentage manganese, and the results were as in Table 2.7.

Null hypothesis H_0: $x_I = x_{II}$
Alternative hypothesis H_1: $x_I \neq x_{II}$

$$\bar{d} = 0.125 \qquad s_d = \sqrt{\frac{\Sigma(d-\bar{d})^2}{n-1}} = 0.20$$

$$s_d/\sqrt{n} = 0.07 \qquad t_e = \frac{0.125}{0.07} = 1.79 \ (\phi = 7)$$

Hence, $0.20 > 2\alpha > 0.10$.

The observed mean difference between the methods could arise purely by chance on more than ten percent of occasions. The difference is not significant, and H_0 is accepted.

Table 2.7 *Results from two analytical methods*

	Method		
Compound	*I*	*II*	$d = x_I - x_{II}$
A	14.7	14.4	0.3
B	10.1	10.0	0.1
C	6.4	6.4	0.0
D	18.3	18.5	−0.2
E	5.4	5.4	0.0
F	13.6	13.2	0.4
G	5.3	5.0	0.3
H	9.3	9.2	0.1

Exercise 2.10 (a)

Suppose that the data in Exercise 2.5 arose as the production (tons/batch) from two different plants operating the same process, where specimens 1–10 were ten different batches of raw material. Is plant 2 significantly superior to plant 1 in its performance?

Answer

$t_e = 4.06 (\phi = 9)$

A one-tailed test is called for. From the results, $0.005 > \alpha > 0.001$. Plant 2 is highly significantly superior.

Exercise 2.10 (b)

A substance C was assessed for its usefulness as an additive to increase the yield of B from A by an otherwise fixed process. To allow for possible variation of A with time, the experiments were carried out in pairs, one with and one without C. The results were as follows. Was the presence of C significant?

	Pair	*1*	*2*	*3*
Yield	*with C*	26.3	25.5	24.8
	without C	25.9	23.6	23.4

Answer

$t_e = 2.79$, $\phi = 2$, $0.10 > \alpha > 0.05$.

This is possibly statistically significant. Further testing is necessary, if the yield increase obtained is economically significant. If not, the process needs further modification.

2.6.4 Non-homogeneous Variances

Equation (2.13) was obtained on the assumption that $s_1 = s_2 = s$, and can be used whenever the ratio of the variances s_1^2, s_2^2 is not significant by F-test (or where long experience suggests that this will be true). Sometimes, however, we wish to compare means where this condition is not fulfilled. For example, we may wish to compare mean results from two analytical methods for the same material. If one method is physical, say, while the other is chemical, the two variances may be significantly different. Significantly different variances might also occur when we are comparing results from experienced and inexperienced workers. For cases where the difference between means is small compared with the overall mean, similarity (homogeneity) of variances can probably be assumed, but (quoting Mead and Curnow[5]) a factor of two for the difference of the means is probably the limit before homogeneity of variance is checked. If there is non-homogeneity, the *t*-test cannot be applied, but we can use the $\tan\theta$ test[6] or the Welch test.[7] The tests are mathematically closely related, being developments of the *t*-test. The Welch test is computationally slightly simpler, and therefore is illustrated here.

Example 2.11: Welch test for difference between two means of different variances

If the experimental value of

$$W = \frac{|\bar{x}_1 - \bar{x}_2|}{\sqrt{\dfrac{s_1^2}{n_1} + \dfrac{s_2^2}{n_2}}} \tag{2.25}$$

(where $\bar{x}_1, \bar{x}_2$ = means, s_1^2, s_2^2 = variances of samples size n_1, n_2) is greater than appropriate table values, then a significant difference between means is indicated. The table values relate to degrees of freedom $(n_1 - 1)$ and $(n_2 - 1)$ and a quantity

$$C = \frac{s_1^2/n_1}{(s_1^2/n_1) + (s_2^2/n_2)} \tag{2.26}$$

Specimen calculation

Is $\bar{x}_1$ significantly greater than $\bar{x}_2$ on the following data?

$\bar{x}_1 = 17.0,\ n_1 = 12,\ s_1^2 = 2.25,\ \phi_1 = 11$
$\bar{x}_2 = 14.5,\ n_2 = 11,\ s_2^2 = 12.1,\ \phi_2 = 10$

Answer

$$s_2^2/s_1^2 = F = 5.38$$

From *F*-tables, single-sided for $\phi = 10, 11$, this *F*-value gives $\alpha \sim 0.005$. The variances are significantly different and cannot be pooled. Hence the *t*-test cannot be used. Instead, compute Welch's statistic *W*:

$$W = \frac{2.50}{1.14} = 2.19$$

Also

$$C = 0.14$$

From tables, for a single-sided test,

	α	0.01		0.05	
	C	0.10	0.20	0.10	0.20
$\phi_1=10,\ \phi_2=10$		2.70	2.63	1.78	1.76
$\phi_1=12,\ \phi_2=10$		2.62	2.57	—	—
$\phi_1=10,\ \phi_2=15$		—	—	1.78	1.76

Therefore, $W_{\mathbf{exptl}}$ is significant at the 5% level: $0.05>\alpha>0.01$.
$\bar{x}_1$ is significantly greater than $\bar{x}_2$.

Exercise 2.11

Is $\bar{x}_1$ significantly different from $\bar{x}_2$ on the following data?

$\bar{x}_1=5.050$, $n_1=16$, $s_1^2=0.0013$, $\phi_1=15$
$\bar{x}_2=5.078$, $n_1=40$, $s_2^2=0.0003$, $\phi_2=39$

Answer

$W=2.98$, $C=0.92$. For a double-sided test, with these degrees of freedom, $W=2.98\rightarrow\alpha<0.02$. The means are significantly different.

2.6.5 The *F*-Test

If we wish to establish that two samples belong to the same population, we should test not only to determine that their means are not significantly different, but also that their standard deviations do not differ significantly. The procedure to examine the latter criterion is the *F*-test, where *F* is the ratio of the two sample variances: see Example 2.12 and Exercise 2.12. In addition, it has already been mentioned that a non-significant value for *F* should be established if the *t*-test is to be used to compare means. Thirdly (and principally) the *F*-test is used to test for significant differences between sources of variation in experimental results, after the total variation in the results has been parcelled out and assigned to these sources by an analysis of variance (ANOVA). This subject is introduced in Section 2.7. The *F*-test and ANOVA are of prime importance in the rest of this book: they form the most mathematically detailed analysis technique whenever there are more than two effects to compare, which happens throughout the statistically designed experiments discussed later.

Example 2.12: to test whether the observed variances of two samples differ significantly

If s_1^2 and s_2^2 are the variances of two sets of observations, then the *F*-test statistic is

$$F=s_1^2/s_2^2 \qquad (2.27)$$

where s_1^2 must always be the larger variance.
Table 2.8 gives critical values (at 1% and 5% probability of error, single-sided) for

Table 2.8 *Some percentage points of the F-distribution for one-tailed tests*

ϕ_2 \ ϕ_1	1	4	7	10
1	161.4	224.6	236.8	241.9
	4052	5625	5928	6056
4	7.71	6.39	6.09	5.96
	21.2	16.0	15.0	14.5
7	5.59	4.12	3.79	3.64
	12.25	7.85	6.99	6.62
10	4.96	3.48	3.14	2.98
	10.04	5.99	5.20	4.85
14	4.60	3.11	2.76	2.60
	8.86	5.04	4.28	3.94
20	4.35	2.87	2.51	2.35
	8.10	4.43	3.70	3.37

The upper line of each pair gives the value of F for $\alpha=0.05$; the lower gives that for $\alpha=0.01$.

the corresponding degrees of freedom ϕ_1 and ϕ_2. ϕ_1, relating to the larger variance, is shown along the top of the table, and ϕ_2 down the left-hand side. If the two sample sizes are n_1, n_2, then $\phi_1=n_1-1$ and $\phi_2=n_2-1$ (but in other applications, ϕ_1 and ϕ_2 can have various values – see, *e.g.* Section 2.7).

Specimen calculation

Consider the mixing of several components in a chemical reactor. To determine the effectiveness of the mixing process at constant stirring rate, eight samples were taken from the reactor after ten minutes and eight further samples after a further ten minutes. The percentages of component C found in each sample were as follows:

10 minutes 38.3 36.6 36.0 33.3 34.8 36.2 37.4 35.4 ($\bar{x}_1=36.0$)
20 minutes 36.2 38.1 34.3 35.4 35.2 37.0 35.2 36.6 ($\bar{x}_2=36.0$)

giving

$$s_1^2=\Sigma(38.3-36.0)^2+\ldots(35.4-36.0)^2/7=2.39$$
$$s_2^2=\Sigma(35.5-36.0)^2+\ldots(36.2-36.0)^2/7=1.48$$

whence

$$F=2.39/1.48=1.61$$

From tables for $F_{7,7}$: α(single-sided) 0.10 0.05
F 2.79 3.79

so $F=1.61$ gives $\alpha>0.10$. The variances are not significantly different. The second period of mixing has not made the bulk significantly more homogeneous.

However, it is important to remember that the results in this Example might have been due not only to lack of uniform mixing but also to variability in extracting and determining C from the samples; or, of course, both.

Exercise 2.12a

Suppose analysis of fifteen individually thoroughly mixed samples of known composition had given the variance due to the analytical method as 0.70. Would the results indicate that the bulk was uniformly mixed after the second mixing?

Answer

$$1.48/0.70 = 2.11$$

From the F-tables, there is more than 10% chance probability of an $F_{7,14}$ value as large as 2.11. Hence the variability in the second set of results could have been caused by the analytical variability, and is not evidence of lack of uniformity in the mixture. It is not evidence of uniformity either: all we can say is that the variability of the results is not significantly greater than the variability of the method.

Exercise 2.12b

If the analytical method could be improved in precision, to give variance $= 0.30$, would it give evidence of non-uniformity after the second mixing?

Answer

Yes $(0.01 > \alpha > 0.001)$.

2.7 TESTING FOR DIFFERENCE BETWEEN SEVERAL SAMPLE MEANS: THE ANALYSIS OF VARIANCE

Testing for significant difference between *several* sample means is done by the F-test, which may be surprising at first sight: how can a test which compares variances he used to compare means? The justification runs as follows:

It can be shown that the means of samples, each of n items, from a single Normal population of variance σ^2, are themselves Normally distributed with variance σ^2/n. If we have m samples, each size n, and their means have a sufficiently greater variance than we would expect from m individuals from a population whose variance is σ^2/n, we conclude that, probably, there is some real physical difference between the means. That is to say, an observed F-value greater than a chosen critical value shows that at least *some* of the sample means do differ at the chosen level of significance.

These m means may have come from m 'treatments', and the term *treatment* is very general indeed. For instance, we could be considering m different analytical methods, m different analysts, m different varieties of wheat in an agricultural test, m different catalysts for a process, m different manufacturers of a raw material, and so on. The variables just given are examples of *qualitative factors*. In other cases, different treatments are, *e.g.*, different temperatures, different pHs, different stirring rates, or different application rates of a given fertilizer. These are examples of *quantitative factors*, variables which can be set at definite numerical values known as treatment *levels*. But the qualitative factors, although they lack numerical settings, can be given levels; arbitrarily we assign, say, four analytical methods as levels 1, 2, 3, 4. By so doing, all variables can be analysed by the same statistical techniques.

2.7.1 Test for Difference of Means between Several Samples of the Same Size (One-way Analysis of Variance)

If we have m treatment levels, and we take n observations at each level, the results are usually tabulated as in Table 2.9, where x_{ij} is the result from the jth observation at the ith level of treatment.

Table 2.9 *Results for n observations on each of m treatments*

Treatment level		*1*	*2*	*3*	...	*m*
Observation	1	x_{11}	x_{21}	x_{31}	...	x_{m1}
number	2	...	...	...	...	...
	.	...	...	...	...	...
	.	...	...	...	...	...
	.	...	...	...	...	...
	n	x_{1n}	x_{2n}	x_{3n}	...	x_{mn}
Totals		Σx_1	Σx_2	Σx_3	...	Σx_m
Mean		$\bar{x}_1$	$\bar{x}_2$	$\bar{x}_3$	...	$\bar{x}_m$

Grand mean $=\bar{x}=(\bar{x}_1+\bar{x}_2+\ldots\bar{x}_m)/m$

The mean of each column is calculated, then the 'grand mean' $\bar{x}$ of these means, and the two variances are calculated as follows:

The observed variance of the treatment means about the grand mean (the *between-means variance*) is given by:

$$s_b^2=\{(\bar{x}_1-\bar{x})^2+(\bar{x}_2-\bar{x})^2+\ldots(\bar{x}_m-\bar{x})^2\}/(m-1) \qquad (2.28)$$

If the null hypothesis H_o is true, s_b^2 is an estimate of σ^2/n, with $m-1$ degrees of freedom, since the standard error of the treatment means is $\sigma/\sqrt{n}$. Therefore, ns_b^2 is an estimate of σ^2, with $\phi=m-1$.

If H_0 is true (*i.e.* the m sets of results come from the same population) the observed variance within a single set is also an estimate of σ^2, with $n-1$ degrees of freedom. We can average these estimates of σ^2 from the m treatments to get an overall estimate of σ^2 which is the *within-means variance*, s_w^2:

$$s_w^2=\{(x_{11}-\bar{x}_1)^2+(x_{12}-\bar{x}_1)^2+\ldots.(x_{mn}-\bar{x}_m)^2\}/m(n-1) \qquad (2.29)$$

We calculate F by

$$F = \frac{ns_b^2}{s_w^2} \tag{2.30}$$

with

$$\phi_1 = m - 1,\ \phi_2 = m(n-1)$$

ns_b^2 represents the variation in results due to the various treatments having different effects. s_w^2 represents the experimental error.

If H_0 is not true, s_w^2 will still be an estimate of σ^2, but ns_b^2 will be increased by the treatment differences. If the experimental F is significantly large, then real difference between the means is probably present: there is evidence of a difference between the treatment effects.

Example 2.13

Suppose we have an original composition (1) and three modifications of it (2, 3, 4) for a cement mix, and four measurements of 24 hour strength are made on each product. Assume the results are as in Table 2.10.

Table 2.10 *24 hour strengths of cement mixtures*

Treatment		1	2	3	4
Observation	1	96	103	111	110
number	2	104	106	113	118
	3	106	105	109	109
	4	102	106	111	115
Means		102	105	111	113

Grand mean $= \bar{x} = 107.75$

Here $m = n = 4$, so we have

$$s_b^2 = (5.75^2 + 2.75^2 + 3.25^2 + 5.25^2)/3 = 26.25$$

and therefore $ns_b^2 = 105.0$.

$$s_w^2 = (6^2 + 2^2 + 4^2 + \ldots\ldots 4^2 + 2^2)/12 = 10.33$$

Therefore

$$F = 105.0/10.33 = 10.16.$$

From tables, $F_{3,12} = 5.95 (\alpha = 0.01)$, $10.80 (\alpha = 0.001)$. So there is highly significant difference between the means.

We can now test the modified-process means against the original-process mean, and against each other, by t-test or similar, to seek which modified process(es) is/are significantly different from the original, and whether there is significant difference between the modifications. If the F-test proves not significant, we can be sure there will not be a significant difference between any pair.

A shorter way of calculating mean squares

Because the total sum of squares of all the deviations $\Sigma(x-\bar{x})^2=\Sigma x^2-(\Sigma x)^2/mn$, it is necessary to calculate only the total of all the results, Σx, the sum of the squared results, Σx^2, and the column totals, Σx_i, proceeding as follows:

(i) Sum all the observations:

$$\Sigma x = x_{11} + x_{12} + \ldots\ldots x_{mn}.$$

(ii) Find a necessary 'correction term' $(\Sigma x)^2/mn$.

(iii) Find the sum of the squared observations:

$$\Sigma x^2 = x_{11}^2 + x_{12}^2 + \ldots\ldots x_{mn}^2.$$

(iv) Obtain the total sum of the squared deviations = (iii) − (ii).

(v) Calculate the between-means sum of squares

$$\frac{\sum_{i=1}^{m} (\Sigma x_i)^2}{n} - \frac{(\Sigma x)^2}{mn}$$

(vi) Then the within-means sum of squares = (iv) − (v).

This is a simple extension of the method illustrated in Example 2.1 for the calculation of a single variance.

Exercise 2.13

Four groups of five testers remove the following numbers of reject items from a production line. Show that there is a significant difference between the groups.

Group	A	B	C	D
	6	9	7	9
	7	10	8	10
	8	10	9	12
	9	12	9	14
	10	14	12	15

Answer

Experimental $F_{3,16}=4.04$; $\alpha \sim 0.025$.

We have compared the variance between means (the numerator of F) with the variance within means (the denominator) and have therefore carried out a simple *analysis of variance*. In Example 2.13 and Exercise 2.13, the observations were divided into m mutually exclusive categories (*i.e.* by the single criterion of different treatments or different groups). We have one set of treatment means, and correspondingly have done a *one-way* analysis of variance. In another case, we may wish to, *e.g.*, obtain results from m analysts using n different analytical methods.

The results will then be divided up by two criteria, with two sets of means (analysts means and methods means) and we have a *two-way* analysis of variance.

In the latter case, if each analyst does each method only once (*i.e.* there are mn observations) the total variance is divided into a between-analysts variance, a between-methods variance, and an error variance. The error variance becomes the denominator of F, and we can analyse for significant effects due to analysts and to methods (see Example 2.14 and Exercise 2.14). In this experimental design, any interaction between the two effects (the differences between analysts in operating a method is not independent of the method being considered) forms part of the error variance. To estimate any such interaction, the experiment must be replicated (Example 2.15 and Exercise 2.15).

2.7.2 Two-way Analysis of Variance on Unreplicated Results

The results are set out as shown in Table 2.11 and the quantities shown in Table 2.12 calculated.

Example 2.14

In order to compare the relative efficiencies of three scrubbing units in removing organic wastes from factory effluent, an observation was made on each unit on each of four successive days. By two-way ANOVA, we can estimate

(i) whether the unit efficiencies differ significantly;

(ii) whether there is significant day-to-day variation (which may be in the effluent, the units, the method of analysis, or some combination)

by comparing the between-units and between-days variances, respectively, with the residual variance. The results were as in Table 2.13. Coding by subtracting 80 gives results as in Table 2.14.

Now, from tables,

	$\alpha = 0.10$	0.05	0.025
$F_{2,6}$	$= 3.46$	5.14	7.26
$F_{3,6}$	$= 3.29$	4.53	6.23

The difference between units is possibly significant ($0.10 > \alpha > 0.05$).

The difference between days is significant ($0.05 > \alpha > 0.025$).

More experiments are desirable, to firm up on whether the units are really different and to investigate the origin(s) of the day-to-day variation.

The limitations of analysing the same data by one-way ANOVA

In Example 2.14, each unit operated once a day. If the twelve experiments were run in random order over the four days, taking no account of how many times a given unit operated in a particular day, the only thing that can be analysed for is the difference between units as compared with fluctuation within units. The figures would be $T = 36$, $T^2/mn = 108$, $\Sigma x^2 = 162$ as before. Therefore, the total sum of squares is unchanged (54) and so is the between-units sum of squares (15.50, with $\phi = 2$). By difference, the within-units sum is 38.50 ($\phi = 9$). The mean squares correspondingly are 7.75 and 4.28, giving experimental $F_{2,9} = 1.81$. Now,

Table 2.11 *Results layout for two-way ANOVA (unreplicated experiment)*

Treatment		1	2	...	m	
						Row sums
Blocks	1	x_{11}	x_{21}	...	x_{m1}	B_1
	2	x_{12}	x_{22}	...	...	B_2
	.	...	...	...	...	..
	.	...	...	...	...	..
	n	x_{1n}	x_{2n}	...	x_{mn}	B_n
Column sums		C_1	C_2	...	C_m	*Grand Total* T

$$T=\sum_{1}^{m} C=\sum_{1}^{n} B$$

Grand mean $= T/mn$
Correction term $= T^2/mn$

Table 2.12 *Two-way analysis of variance table (unreplicated experiment)*

Source of variation	*Sum of squares*	*Degrees of freedom,* ϕ	*Mean square*
Between treatments (columns)	$\dfrac{\left(\sum_{1}^{m} C^2\right)}{n}-\dfrac{T^2}{mn}$	$m-1$	$\dfrac{\text{Sum of squares}}{m-1}=MS_C$
Between blocks (rows)	$\dfrac{\left(\sum_{1}^{n} B^2\right)}{m}-\dfrac{T^2}{mn}$	$n-1$	$\dfrac{\text{Sum of squares}}{n-1}=MS_B$
Residual	see below	$(m-1)(n-1)$	$\dfrac{\text{Sum of squares}}{(m-1)(n-1)}=MS_R$
Total	$\sum_{1}^{m}\sum_{1}^{n} x^2-\dfrac{T^2}{mn}$	$mn-1$	

Residual sum of squares, MS_R = total sum − sum between treatments − sum between blocks

Table 2.13 *Percentage removal of organics by three scrubber units*

Units	A	B	C
Day 1	87	86	82
2	83	81	80
3	86	83	83
4	82	81	82

Table 2.14 *Two-way analysis of variance of scrubber efficiency*

Unit		A	B	C	*Block totals, B*	B^2
Day	1	7	6	2	15	225
	2	3	1	0	4	16
	3	6	3	3	12	144
	4	2	1	2	5	25
Column totals	C	18	11	7	*Grand total* $T=36$	
	C^2	324	121	49		

$\Sigma B^2=410$; $\Sigma C^2=494$; $T^2=1296$
$\Sigma\Sigma x^2=7^2+3^2+\ldots.2^2=162$; $m=3$; $n=4$

Source of variation	*S.o.S.*	ϕ	*MS*	*F*
Between-units	$\frac{494}{4}-\frac{1296}{12}$ $=15.50$	2	7.75	$\frac{7.75}{1.64}=4.73$
Between-days	$\frac{410}{3}-\frac{1296}{12}$ $=28.67$		39.56	$\frac{9.56}{1.64}=5.83$
Residual	$54.00-15.50-28.67$ $=9.83$	6	1.64	
Total	$162-108=54.00$	11		

from tables, $F_{2,9}=3.01$ for $\alpha=0.10$. The difference between the units is now *not* significant. We see that blocking the experiment gives more indication of significance. In general, for blocked and unblocked experiments with the same number of observations in each, the former gives more information: it analyses for two (or more) sources of information instead of only one. It is also more sensitive and thus more likely to yield significant information. Hence Box, Hunter, and Hunter's advice to 'block what you can and randomise what you cannot'.[8]

Exercise 2.14

Five research assistants tried out three different methods to synthesize a particular product. Table 2.15 shows the parts per thousand of a damaging impurity after one recrystallization. Is there significant difference between the methods or between the assistants?

Answer

Mean squares: between-methods 64.85, between-assistants 138.4, residual 28.45.

Table 2.15 *Parts per thousand impurity*

Method	A	B	C
Assistant V	35	25	18
W	30	20	20
X	22	29	15
Y	16	7	13
Z	10	13	11

$F_{\text{between methods}} = 2.28$ ($\phi = 2, 8$): not significant at $\alpha = 0.10$
$F_{\text{between assistants}} = 4.86$ ($\phi = 4, 8$): significant at $\alpha = 0.05$.

2.8 RANDOMIZATION AND BLOCKING

It is essential that any set of related experiments be done in random order. For instance, in an investigation, the three treatments examined might be three batches of raw material. It is essential that we do *not* do all the batch 1 experiments first, then those with batch 2, and finally those with batch 3. If we do, and some reagent which we believe to be stable is in fact deteriorating, we will ascribe the worsening results, incorrectly, to the raw material. We are confused; in statisticians' language, we have confounded the difference between the batches with the change in the reagent. To avoid this, we must mix up the three experiments randomly through time. This principle, *randomization*, has been called the most important principle of good experimentation.[9] It minimizes the chance of an unidentified systematic error vitiating the results.

But the experimenter may be aware of an unwanted but unavoidable source of variability. For instance, it may be inevitable that studying variations of a chemical process has to be carried out with several, different, batches of material. If so, the results on any one batch will tend to be more similar (have less random error) than those on different batches. Because we wish to minimize our random (error) variance, to make our F-test sensitive, it is sensible to ensure that each variation of the process is studied with each batch of material. We therefore divide our work up into blocks, each block representing a batch. If an equal number of measurements is made on each treatment in a block, and if the order of tests within a block is random, we have a *randomized block design*. Example 2.14 and Exercise 2.14 use such a design without replication.

Randomization and blocking are valuable devices for dealing with (unknown and known) *unavoidable* sources of variability. Comparing several treatments within a block is an extension of the paired comparisons technique described in Section 2.5.5, and similarly it increases the likelihood of identifying significant effects, as we saw in Example 2.14.

Identifiable *avoidable* sources of variability should of course be minimized, or this likelihood will be reduced.

One such avoidable source of variation is the passage of time. Suppose we can carry out three unit experiments on each block in a day. A randomized block

Table 2.16a *Randomized block design*

Batch		*Run order*	
1	A	B	C
2	A	C	B
3	C	A	B

Table 2.16b *Latin square design*

Batch		*Run order*	
1	A	B	C
2	C	A	B
3	B	C	A

Table 2.16c *Graeco-Latin square*

Batch		*Run order*	
1	Ax	By	Cz
2	Cy	Az	Bx
3	Bz	Cx	Ay

design may result in the run orders shown in Table 2.16a for studying the effect of three catalysts A, B, C on a reaction using three batches of starting material. A tends to be used earlier in the day, and B later.

If there is some trend connected to time of day, it may be confounded with the difference between A and B.

The way out of the difficulty is to use a *Latin Square* design, in which each catalyst is used once on each batch and also appears once in each time position (Table 2.16b). In a randomized block design, there is only one *blocking variable* (batch in Table 2.16a). This simply ensures that each treatment (catalyst) is applied to each batch. In a Latin Square, there are two such variables (batch and run order).

A full factorial for three batches, three catalysts, and three run orders would require a 3^3 design: 27 treatment combinations. This would estimate the three main effects and all the interactions. The Latin Square needs only nine runs, but estimates main effects only. It is thus suitable if one can (or does) assume that interactions are insignificant.

If we wish to investigate using three blocking variables, a *Graeco-Latin Square* is used. For instance, if the previous investigation was carried out by three operators x, y, z, we would have a design as in Table 2.16c. The extension of these designs to more variables, and the analyses of variance, can be found in standard textbooks.[10,11] Sometimes the number of treatments to be examined exceeds the most suitable block sizes; we might, for instance, wish to examine five catalysts but can do only three experiments per day. In such circumstances, *Youden* and other *balanced incomplete block* designs are available.[12,13]

2.9 TWO-WAY ANALYSIS OF VARIANCE ON REPLICATED RESULTS

In the two-way analysis of variance discussed in Section 2.7.2, we cannot assess the extent to which a non-zero residual sum of squares is caused by interaction between the two effects, rather than by random error. In order to estimate interaction and random effects separately, each unit experiment (each treatment combination) must be replicated. In addition to randomization, *replication* has been described as a basic statistical requirement for a good experiment,[14] and the replication must repeat every detail if the effects estimates are to be accurate: use the same solutions, equipment, procedures and so on.

The two-way ANOVA table for replicated results is the same as for unreplicated results in respect of the total, between-columns, and between-rows sums of squares, *except that* every divisor is multiplied by r, the degree of replication. The residual variation, due to random error, is estimated from the within-replicates variation. The residual sum of squares is given by

$$\sum_{i=1}^{m}\sum_{j=1}^{n}\sum_{k=1}^{r}(x_{ijk}-\bar{x}_{ij})^2/r-1 \tag{2.31}$$

where x_{ijk} is the kth replicate result in the ith row and the jth column, and $\bar{x}_{ij}$ is the mean result in the ith row and the jth column.

The interaction sum of squares is then obtained by difference:

total − between columns − between rows − residual
= interaction sum of squares.

The degrees of freedom, ϕ, are as follows:

between columns (m columns)	$m-1$
between rows (n rows)	$n-1$
total (m columns, n rows, r replicates)	$mnr-1$

ϕ for an interaction is the product of the ϕ values for the interacting factors; hence:

interaction	$(m-1)(n-1)$

By difference,

residual	$mn(r-1)$

The ANOVA table therefore is as shown in Table 2.17.

Example 2.15

Imagine an investigation of electric resistance welding at four current densities and three applied pressures, and with duplicate runs. Then assume that the quality of the resultant welds is measured on some arbitrary scale, with results as in Tables 2.18 (a) and (b).

The analysis of variance for the duplicated results is therefore as shown in Table 2.19. The F-values, respectively, can be compared with $F_{3,12}=10.8$ at $\alpha=0.001$,

Table 2.17 *Tow-way ANOVA table for replicated results*

Source of variation	*Sum o squares*	*Degrees of freedom*	*Mean square*
Between columns	$\frac{\Sigma C^2}{nr}-\frac{T^2}{mnr}$	$m-1$	S.o.S./$m-1$
Between rows	$\frac{\Sigma B^2}{mr}-\frac{T^2}{mnr}$	$n-1$	S.o.S./$n-1$
Interaction	by subtraction	$(m-1)(n-1)$	$\frac{\text{S.o.S.}}{(m-1)(n-1)}$
Residual	$\frac{\Sigma(x_{ijk}-\bar{x}_{ij})^2}{r-1}$	$mn(r-1)$	$\frac{\text{S.o.S.}}{mn(r-1)}$
Total	$\Sigma(x_{ijk})^2-\frac{T^2}{mnr}$	$mnr-1$	

ΣC^2 = sum of squares of column sums. ΣB^2 = sum of squares of row sums. m = number of columns. n = number of rows. r = number of replicates. $\Sigma(x_{ijk})^2$ = sum of squares of individual results. T = sum of individual results.

Table 2.18 *Duplicate resistance welding results*

(a) *Current densities*	1	2	3	4
Pressures 1	5, 4	7, 8	10, 9	9, 7
2	3, 4	4, 5	6, 8	8, 8
3	4, 2	5, 5	7, 6	10, 9

Summing duplicates gives

(b) *Current densities*	1	2	3	4	*Row sums B*
Pressures 1	9	15	19	16	59
2	7	9	14	16	46
3	6	10	13	19	48
Columns sums C	22	34	46	51	*Total* $T=153$

$\Sigma C^2=6357$ $\Sigma B^2=7901$ $T^2/mnr=23\,409/24=975.38$

$F_{2,12}=6.93$ at $\alpha=0.01$, and $F_{6,12}=3.00$ at $\alpha=0.05$. Hence the difference between current densities is very highly significant, the difference between pressures is significant, and the interaction is possibly significant.

Exercise 2.15

Examine the data in Table 2.20 for significant effects, if any, arising from three different alloy compositions A, B, C and three different heat treatments I, II, III.

Table 2.19 *Analysis of variance of duplicate welding results*

Source of variation	*Sum of squares*	*Degrees of freedom*	*Mean square*	*F*
Between columns	$\frac{6357}{6} - 975.38$	3	28.04	35.5
Between rows	$\frac{7901}{8} - 975.38$	2	6.13	7.76
Interaction	by difference, 13.75	6	2.29	2.90
Residual	$(0.5^2 + 0.5^2 + \ldots 0.5^2)/1$ $= 9.50$	12	0.79	—
Total	$(5^2 + 4^2 + \ldots 9^2) - 975.38$ $= 119.62$	23		

Table 2.20 *Tensile strength of light alloys*

Composition		A	B	C
Heat treatment	I	24.2, 23.1	27.7, 26.8	28.4, 29.8
	II	25.6, 23.2,	27.1, 26.8	27.0, 27.8
	III	23.9, 24.6	25.6, 25.9	26.4, 27.4

Answer

Mean squares: composition 21.52, treatment 1.62, interaction 1.31, residual 0.67; $F_{2,9}$ (composition) 32.12 (very highly significant), $F_{2,9}$ (treatment) 2.42: $\alpha \rightarrow 0.10$ (might justify further experiments), $F_{4,9}$ (interaction) 1.96 (not significant).

2.10 NECESSARY DEGREE OF REPLICATION

The reader may very well ask 'How many replications will be required in a given situation?' This is a question to which there is no simple, clear-cut answer. Replication makes a given difference between means (a given effect) easier to detect at a given level of significance. So the first thing to decide is: what size of difference is the experiment required to detect? The greater the degree of replication, the more the experiment will be capable of detecting smaller and smaller differences. This can be overdone, until an effect which is for all practical purposes insignificant will be declared statistically significant. This *distinction between statistical and practical significance* should always be remembered, and not only in this context of forcing significance by repeated testing. A change may be statistically significant and yet impracticable on grounds of cost, convenience, or other factors or simply too small to be worthwhile. Hence, statisticians emphasize the importance of *estimation* (the statement of means and associated confidence intervals) as well as significance testing.

It is clear that the degree of replication should be determined by the minimum required detectable difference. Mead and Curnow,[15] following Cochran and Cox,[16] proceed as follows. Suppose the experimenter wishes to detect any difference between the effects of two treatments as large as d units of yield. Suppose also that experience with similar experimental studies suggests that the variance per replication is some value σ^2. Then, for r replications, the standard error of the difference between treatment means should be $\sqrt{2\sigma^2/r}$.

Now an estimated difference between means equal to twice this standard error should be significant at about the 5% level. But, of course, our estimate of the true difference may be low. To allow for this, a reasonable rule is to say that an estimated difference equal to *three* times the standard error will five times out of six be significant at the 5% level. Hence, if we want this degree of confidence to continue with our work, we must operate with such a within-block variance σ^2 and degree of replication r that the observed difference between treatments equals $3\sqrt{2\sigma^2/r}$.

Suppose the minimum difference which will persuade us to go on is $0.2m$, where m is the overall mean result over both treatments (approximately, we want at least a 20% improvement). Then we must have $0.2m = 3\sqrt{2\sigma^2/r}$. Re-arranging gives $r = 450\sigma^2/m^2$.

If $\sigma = 0.1m$, we need $r = 4.5$: *i.e.* we must run the whole experiment five times. If $\sigma = 0.06m$, we need $r = 1.62$: only duplication is required. This large reduction in replication once again shows the importance of reducing experimental variability: find a good technique, practise it until proficient, and stick to it like glue!

2.11 BENEFITS OF SEQUENTIAL EXPERIMENTATION

An excellent way of seeking significance, without over-striving for it, is to carry out the experiment sequentially. If a factorial or similar design is being done, run through the design and analyse the results; run through it again and analyse the cumulative results (by which time the duplication allows interactions to be estimated) and run through again as necessary. Each extra run provides extra degrees of freedom for the error variance, making finding significance easier; but unnecessary runs (or runs which would provide spurious significance) are avoided. Sequential testing procedures, using only one unit experiment at a time, are described in Section 3.2.3. Sequential procedures also have advantages in separating main effects from interactions and estimating the first-order interactions of significant factors (Chapter 7) and in optimizing procedures (Chapters 9 and 11).

2.12 OTHER TESTS OF SIGNIFICANCE

There are other very useful tests of significance, based on the Normal distribution or on distributions related to it. Simple graphical methods are discussed in Chapter 4 and Section 5.7. Other methods, which are not discussed in this book but can be found in standard texts on statistical analysis, include the *binomial test of proportions* used to test, *e.g.*, for difference in percentage defectives from two production lines, and the *chi-squared test*, which applies to two or more proportions and also to many kinds of data which fall or can be divided into categories (*e.g.* numbers of radioactive emissions from a given source in successive unit intervals of time, the number of defective production items from different groups of workers, or the numbers of students awarded first, second, or third class honours degrees).

In addition, there are numerous *non-parametric* (otherwise known as *distribution-free*) tests, which make no assumptions about the distribution of the observations. They are based on arranging the data in order or ranking them in some other way, and are computationally simple. However, this may be thought relatively unimportant in the presence of calculators and computers, and they have lower power (see Section 3.3) than parametric procedures such as the t- and F-tests. Box, Hunter, and Hunter[17] have concluded only that they are 'occasionally useful in the analysis of highly non-Normal data'. This is particularly the case where analysis can be based on the means of relatively large samples (*e.g.* $n > 30$) as the *central limit theorem* of statistics states that the distribution of such means is approximately Normal irrespective of the distribution of the individual values.

REFERENCES

1. A.S.C. Ehrenberg, 'A Primer in Data Reduction', Wiley, London, 1984.
2. 'Statistical Methods in Research and Production', ed. O.L. Davies and P.L. Goldsmith, 4th revised Edn., Longman, London, 1978.
3. J.B. Kennedy and A.M. Neville, 'Basic Statistical Methods for Scientists and Engineers', 2nd Edn., Harper and Row, New York, 1976.
4. C. Chatfield, 'Statistics for Technology', 3rd revised Edn., Chapman and Hall, London, 1989.
5. R. Mead and R.N. Curnow, 'Statistical Methods in Agricultural and Experimental Biology', Chapman and Hall, London, 1983, pp. 109–111.
6. R.A. Fisher and F. Yates, 'Statistical Tables for Biological, Agricultural and Medical Research', 6th Edn., Longman, Harlow, 1974.
7. W.H. Trickett, B.L. Welch, and G.S. James, *Biometrika*, 1956, **43**, 203–205.
8. G.E.P. Box, W.G. Hunter, and J.S. Hunter, 'Statistics for Experimenters', Wiley, New York, 1978, p. 103.
9. Reference 4, p. 230.
10. 'The Design and Analysis of Industrial Experiments', ed. O.L. Davies, 4th Edn., Longman, London, 1978, pp. 159ff.
11. Reference 8, Chapter 8.
12. D.R. Cox, 'Planning of Experiments', Wiley, New York, 1958, Chapter 11.
13. Reference 10, Chapter 6.

14. Reference 5, p. 33.
15. Reference 5, p. 285.
16. W.G. Cochran and G.M. Cox, 'Experimental Designs', 2nd Edn., Wiley, New York, 1957.
17. Reference 8, p. 82.

CHAPTER 3

Some Important Concepts in Design and Analysis

The importance of randomization and blocking (to take account of unavoidable sources of variability) and of replication (to give estimates of interaction effects and of experimental error) has been emphasized in Section 2.8 and 2.9. Some additional important principles are dealt with in this chapter.

3.1 CONFIDENCE LIMITS AND SIGNIFICANCE TESTS

Chapter 2 was concerned with two processes: that of *estimation* (the use of experimental data to quote a population mean and its associated confidence limits) and that of *significance testing* (determining the probability that a difference between a sample mean and some other value is due only to random error). The explanation I have thought necessary has given more space to the latter than to the former; but that does not mean that significance testing is more important than estimation. Statisticians, in fact, are sometimes critical of non-statisticians' over-emphasis on significance tests.

For example, 'I am disturbed at the present tendency of non-statisticians to overdo significance tests . . . there is no point in testing null hypotheses which are obviously untrue . . . (or which are) obviously going to be accepted . . .. In some journals . . . it has become virtually impossible to present any results without carrying out a test and giving a p value. . . .. *It should be realised that in general (though not always), estimation is more important than significance testing*'.[1] Box, Hunter, and Hunter also emphasize the importance of confidence limits with the statement 'Significance testing . . . has been . . . greatly overworked, and in many cases where significance statements have been made, it would have been better to provide an interval within which the value of the parameter would be expected to lie'.[2]

The importance of calculating the confidence interval is illustrated by the following example. Two chemical plants produced a particular solution, with results as follows:

Plant	*Mean % active agent*	*Standard deviation*	*Number of batches*	*Standard error of mean, SE*
A	28.0	1.50	5	$1.5\sqrt{5}$
B	29.5	1.29	8	$1.39\sqrt{8}$

$$\text{Standard error of difference between means} = \sqrt{\frac{1.5^2}{5} + \frac{1.39^2}{8}}$$

$$= 0.831$$

Therefore $t_e = 1.5/0.831 = 1.81$ ($\phi = 4 + 7 = 11$).
From the t-table, for a two-tailed test with $\phi = 11$, $t = 1.81 \rightarrow \alpha \sim 0.10$.
Thus the means are judged unlikely to be significantly different.

But, from equation (2.11), the *confidence limits of the difference* are expressed by

$$(\bar{x}_B - \bar{x}_A) \pm t_{table}(SE) \tag{3.1}$$

and, because $t_{\varphi=11} = 1.80$, from the table, for a two-tailed test with $2\alpha = 0.10$,

the 90% confidence limits are $1.5 \pm 1.80(0.831) = 0 \rightarrow 3.0$.

There is thus a 5% chance that the true $(\bar{x}_B - \bar{x}_A)$ = zero or less, and a 5% chance that it is 3.0 or more.

If a difference of 3.0% in the true means would be technologically or economically significant, there is a 1 in 20 chance that it exists. Extra testing is desirable under such circumstances, and this would not be apparent from significance testing alone. The two calculations are of course related (see Section 2.5.4 in respect of this example).

3.2 DETERMINATION OF THE NECESSARY DEGREE OF TESTING, AND CRITERIA FOR ACCEPTANCE

The phrase 'extra testing is desirable' prompts the question as to how much testing needs to be done in a given situation. The answer to this depends on setting probability limits for two *errors of inference* which may occur when we analyse results. Suppose that our null hypothesis H_0 is that there is no difference between two means. Then,

(i) we may reject H_0 when in fact it is true (known as a *Type I* or *α-error*, or *error of the first kind*).

(ii) we may accept H_0 when in fact it is not true (a *Type II* or *β-error*, or *error of the second kind*).

We want a situation in which a difference of practical importance is nearly always detected (β-errors are rare) and where unimportant differences are rarely classified as significant (α-errors are rare). It turns out from the theory of the Normal distribution that we can control both types of error by choosing the number of experiments to be done. There are two ways of choosing: by prior calculation or by sequential testing. For prior calculation, we proceed as in Section 3.2.1. It also follows from the Normal distribution that we can calculate an experimental mean which must be achieved if a product is to be deemed acceptable (see Section 3.2.2). Sequential techniques, which are often more efficient, are dealt with in Section 3.2.3.

3.2.1 Comparison of an Experimental Mean $\bar{x}$ with a Standard Value μ, and the Comparison of Two Means $\bar{x}_1$ and $\bar{x}_2$

It can be shown that, to compare $\bar{x}$ and μ (which may be, *e.g.*, the specification mean for a product, or the mean value for numerous previous batches) we must make one measurement on each of n samples, where n is defined by

$$n = (U_\alpha + U_\beta)^2 \frac{\sigma^2}{d^2} \tag{3.2}$$

In equation (3.2), U_α and U_β are the values of the standard Normal deviate for the chosen values of α and β. The equation shows that n is reduced if

(i) the total variance of the process (from raw materials to final analysis) is reduced,
(ii) the difference $d = \bar{x} - \mu$ is widened (the product specification is relaxed),
(iii) one or both of α and β is/are increased (the chance of erroneous conclusions is allowed to increase).

It is clear from this that the only desirable way of reducing the cost of testing is (i), the reduction of variability in the total process.

The use of equation (3.2) is illustrated in Example 3.1.

Example 3.1

To ensure satisfactory packing of a powder in containers, it is desirable that its bulk density, x, shall be such that

(i) if $\bar{x} = 2.40$, there shall be 99% chance of correctly rejecting the material ($\alpha = 0.01$)
(ii) if $\bar{x} = 2.48$, there shall be 95% chance of correctly accepting the material ($\beta = 0.05$).

Many previous batches have shown that the standard deviation $\sigma = 0.06$. The criteria for acceptance and rejection imply that there is some value of $\bar{x}$ which must be exceeded for acceptability: this is a one-tailed test. For $\alpha = 0.01$, $\beta = 0.05$, $U_\alpha = 2.33$, $U_\beta = 1.64$. Therefore, $n = (2.33 + 1.64)^2 (0.06)^2 / (0.08)^2 = 8.87$. Hence it is necessary to make one measurement on each of nine samples.

This is very likely to be too many samples in practice. But setting both α and β at 5% only reduces the number of samples to seven. However, if the standard deviation of the overall process can be halved, $n = 2.22$: we need take only three samples. Reduction of variability cannot be over-stressed. It produces a more-reliable product, reduces mixing-off and recycling costs, and reduces the cost of testing. (Equally, of course, the production mean must also be near the specification mean – minimal systematic error – for there is no point in systematically producing 'off-spec' goods.)

If *two experimental means* are to be tested against one another, $2n$ tests must be done to estimate each mean:[3]

$$2n = 2\left\{[U_\alpha + U_\beta] \frac{\sigma}{d}\right\}^2 \tag{3.3}$$

If the two means have significantly different standard deviations s_1 and s_2, the equation is

$$2n=2\left\{\frac{[U_\alpha+U_\beta][0.5\sigma_1^2+0.5\sigma_2^2]^{\frac{1}{2}}}{d}\right\}^2 \tag{3.4}$$

If a two-tailed test is necessary (for instance, if in Example 3.1 the powder could be produced either too dense or too fluffy), substitute $U_{0.5\alpha}$ and $U_{0.5\beta}$ in equations (3.2), (3.3), and (3.4), where α and β are the total, two-tailed probabilities. Hence, for $\alpha=0.01$, $\beta=0.05$, $U_{0.5\alpha}=2.58$, $U_{0.5\beta}=1.96$, and n in Exercise 2.16 would increase from 8.87 to 11.59: twelve tests would be necessary. But, once again, if the overall standard deviation can be halved, the testing can be reduced by three-quarters.

These equations assume that σ is known (at least sufficiently accurately to give approximately the right number of tests). If this is not the case, a small pilot experiment may give an estimate of σ which can be used to plan a larger experiment. Alternatively, an estimate of σ may be available from earlier experiments. Under these circumstances, the *t*-test should replace the Normal test. The *t*-test has lower power than the Normal test (see Section 3.3), and calculations by Cooper[4] indicate that, for samples of ten observations, use of the *t*-test is about 5% less likely to reject a false null hypothesis H_0 (and correspondingly about 5% less likely to accept a true H_1 of significant difference) than is the Normal test. O.L. Davies and his colleagues produced tables for the numbers of observations for the *t*-test situation,[5] and Davies and Goldsmith devoted Chapter 5 of their textbook to choosing the number of observations.[6]

3.2.2 Criterion $\bar{x}^*$ for Acceptance of a Batch

It follows from equation (2.4b) that there is a value $\bar{x}^*$ which the experimental mean $\bar{x}$ must exceed if the material in Example 3.1 is to be accepted:

$$\bar{x}^*=\mu_0+U\sigma/\sqrt{n} \tag{3.5}$$

If we set μ_0 equal to the value $\bar{x}=2.40$ corresponding to $\alpha=0.01$, then $\bar{x}^*$ is the upper confidence limit for material with a true population mean$=2.40$. There is therefore less than 1% chance that an observed sample mean $\bar{x}>\bar{x}^*$ will have come from a population mean$=2.40$. Using equation (3.5) with $\sigma=0.06$, $n=9$ gives

$$\bar{x}^*=2.40+2.33(0.06)/\sqrt{9}=2.447.$$

If we accept only batches with mean density $\geqslant 2.447$, we shall have achieved our aim of accepting only those batches with mean bulk density >2.40 at the 1% level and <2.48 at the 5% level.

Exercise 3.1

It is also possible to work down from the upper limit, using the equation $\bar{x}^*=\bar{x}_\beta-U_\beta\sigma/\sqrt{n}$, where $\bar{x}_\beta$ is the value of $\bar{x}$ corresponding to the set β value. Show, for $\bar{x}_\beta = 2.48$, $\beta = 0.05$, $\sigma = 0.03$, $n = 3$, that $\bar{x}^* = 2.452$. (Note that working from $\bar{x}_\alpha = 2.40$, $\alpha = 0.01$, $\sigma = 0.03$, $n = 3$ would give $\bar{x}^* = 2.440$. To give equality of the two values for $\bar{x}^*$ would require the impossible situation of 2.22 tests.)

If a two-tailed test is required, this is equivalent to modifying Example 3.1 to, say,

(i) if $\bar{x} = 2.40$, there shall be 99% chance of rejection ($\alpha_{\text{lower}} = 0.01$)
(ii) if $\bar{x} = 2.56$, there shall be 99% chance of rejection ($\alpha_{\text{upper}} = 0.01$).

There are then an upper and a lower limit on $\bar{x}^*$:

$$\bar{x}^*_{\text{lower}} = 2.40 + 2.58(0.06)/\sqrt{12} = 2.445$$
$$\bar{x}^*_{\text{upper}} = 2.56 - 2.58(0.06)/\sqrt{12} = 2.515$$

We can check that these limits correspond to 95% probability of correct acceptance at $\bar{x} = 2.48$, as before, by using $\beta = 0.05$, $U_{0.5\beta} = 1.96$, $\sigma = 0.06$, $n = 12$: the limits are calculated as $2.48 \pm 1.96(0.06)/\sqrt{12} = 2.446 \rightarrow 2.514$.

3.2.3 Sequential Testing

Where practicable, it is usually (not always) more efficient to test sequentially: the number of experiments necessary is determined by the results themselves. If the first few results appear very conclusive, little further work need be done; if the decision is still in the balance, more work is required. If the time needed to obtain test results is short compared with the length of a full experimental run (for instance, a rapid instrumental analysis after the long synthesis which preceded it) it is not necessary to plan the total experiment in advance. Even in cases where we are using highly organized procedures such as factorial experiments, we can plan, carry out, and analyse the work in sections (see Chapters 6 and 7).

In *sequential testing* it is often convenient to plot results, in an appropriate form, against the number of experiments. As each result is recorded on the plot, either

(i) stop testing because a conclusion can be reached (a hypothesis is accepted or rejected), or
(ii) continue testing; the decision is still uncertain.

Suppose we wish to know if the average yield of some process is significantly greater than some standard value μ. On the vertical axis we plot $\Sigma(x_i - \mu)$, the cumulative sum to date of deviations of the individual yields from μ. The horizontal axis represents the number of results to date, $1, 2 \ldots i \ldots$ On the graph (Figure 3.1) are two parallel lines, B_1 and B_2, the positions of which depend on the chosen values of α and β, the standard deviation σ, and the difference it is important to detect, d. The intercepts on the vertical axis, h_0 and h_1, are given by

$$h_0 = -\log_e\left(\frac{1-\alpha}{\beta}\right)\frac{\sigma^2}{d} \qquad (3.6)$$

$$h_1 = +\log_e\left(\frac{1-\beta}{\alpha}\right)\frac{\sigma^2}{d} \qquad (3.7)$$

The gradient of B_1 and $B_2 = 0.5d$.

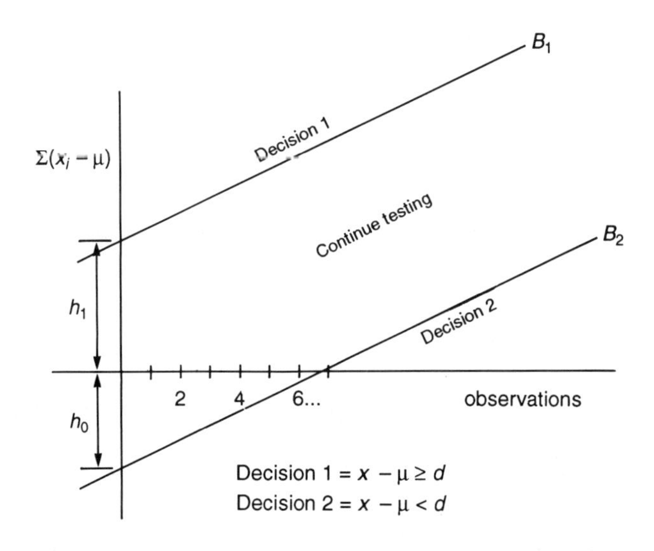

Figure 3.1 *Sequential testing*

If the most-recent point on the graph lies between B_1 and B_2, testing is continued. Above B_1, we decide that $\bar{x}$ is significantly greater than μ. Below B_2, we decide there is no difference as large as d. Either way, as soon as a point plots outside B_1 or B_2, we stop testing.

The values of points on lines B_1 and B_2 are given by

$$B_1 = h_1 + 0.5nd \tag{3.8}$$

$$B_2 = h_0 + 0.5nd \tag{3.9}$$

where $n = 0,1,2\ldots$ tests.

There is no need actually to prepare a graph; these values can simply be tabulated, and cumulative values of $\Sigma(x_i - \mu)$ compared with them by inspection.

The importance of working with high precision (small σ) and of setting realistic values for d, α, and β is exemplified by equations (3.6) and (3.7):

(i) if σ is large, B_1 and B_2 are far apart: it takes a long time to make a decision;
(ii) if d is small, the effect is the same as in (i);
(iii) if α is small, h_1 is large: the plot is less likely to rise above B_1 and we are less likely to accept a significant difference;
(iv) if β is small, the plot is less likely to fall below B_2, and we are less likely to accept no significant difference.

In both (iii) and (iv), the number of tests tends to be greater, but we are less likely to commit (respectively) an α- or β-error.

To review the concepts of Section 3.2 and compare them with an unplanned experiment, work through Exercise 3.2.

Exercise 3.2

The following thirteen results (% purity) were obtained on a single batch, in the order shown:

91.2, 90.8, 91.8, 92.4, 91.4, 91.0, 90.8, 92.0, 90.2, 91.0, 91.4, 92.2, 92.0.

(i) Carry out a t-test to see whether the mean is significantly greater than 90.0. If so, accept the batch.

(ii) If the number of experiments had been planned in advance under the following assumptions, how many measurements would have been required? Assumptions: population standard deviation = 1.0; chance of accepting a batch if the true result is 90.0 or less = 1%; chance of rejecting a batch if the true result is 92.0 or more = 5%. If the tests done were those at the beginning of the list above, what would the decision have been?

(iii) If the experiment had been done sequentially under the assumptions of part (ii), and the tests done were again those at the beginning of the list, how many measurements would have been made, and what would the decision have been?

Answer

(i) $\bar{x}=91.4$, $s/\sqrt{n}=0.18$, $t=7.78$ ($\phi=12$). $\bar{x}$ is very significantly higher than 90.0. Accept the batch.

(ii) $n=4$. $\bar{x}^*$ for acceptance $=90.0+2.33/\sqrt{4}=91.17$. $\bar{x}$ for the first four tests $=91.55$. Accept the batch.

(iii) $n=5$, when $\Sigma(x_i-\mu)$ rises above B_1. Accept the batch. Note that here the sequential approach takes one more test than the planned approach. This is because of the magnitudes of the first five results. If the last three results shown had been obtained first, they would have been sufficient for acceptance, as B_1 values from equation (3.8) and the experimental results show:

	No. of tests	0	1	2	3	4	5
	B_1 *values*	2.28	3.28	4.28	[5.28]	6.28	[7.28]
$\Sigma(x_i-\mu)$	*first five*	0	1.2	2.0	3.8	6.2	[7.6]
	last three	0	1.4	3.6	[5.6]	—	—

The number of required observations is usually less by the sequential approach.[7]

3.2.4 The Advantage of Specifying Both α- and β-Values

The t-value obtained in Exercise 3.2 corresponds to $\alpha \ll 0.001$, and such an unnecessarily high value indicates over-testing. If this had not been indulged in, the economy of the planned or sequential approach would have been less. In fact, the first four results listed would have given $t=4.43$ ($\phi=3$) corresponding to $0.025>\alpha>0.01$, quite sufficient to indicate significant departure from the null hypothesis $\mu=90.0$. But *a significance test guards only against α-error*: we have no idea how much above 90.0 the population mean is. There is less than 2.5% chance that the material is less than 90% pure (the minimum acceptable) so our customer is unlikely to complain. But we may be giving away material much above requirements.

By specifying both α- and β-values, we can do better than that. For the same (or about the same) work by the planned or sequential approach, we can be 99% sure that the material is not below the minimum specification (90% pure) and 95% sure it is not above the maximum (92% pure).

3.3 STATISTICAL TEST POWER

We saw in Section 3.2 that in quality control the necessary amount of testing and the criteria for acceptance are partly governed by the magnitudes of α- and β-uncertainty we allow ourselves. The *power* of a statistical test is defined as the probability of rejecting a false null hypothesis H_0; or equivalently, the probability of accepting a true alternative hypothesis H_1. It equals $1-\beta$, where β = fractional probability of committing the β-error (accepting a false H_0). It would therefore seem that if, in any given case, we define β, then the power is fixed. This, however, is not correct. Firstly, the likelihood of β-error decreases if we increase α (we make it more likely that we will accept H_1, and thus we are less likely to accept H_0). Hence $1-\beta$ increases with increase of α. Secondly, power increases with sample size n (smaller differences such as $|\bar{x}-\mu|/\sigma$ are needed for significance; hence $|\bar{x}-\mu|=0$ is more easily rejected). Thirdly, for given n, larger values for $|\bar{x}-\mu|$ lead to easier rejection. Fourthly, a one-tailed test more easily finds significance: it is more powerful than its two-tailed equivalent. The actual *operational power* (discriminating power) of a given test is therefore largely at the analyst's command, for he or she can vary the first, second, and fourth items.

Nonetheless, if these variables are held constant, different statistical tests do have different powers. The *power function* expresses the power of a test as the probability of detecting a change in result as a function of that change. More mathematically, it is the relationship between the probability of rejecting the null hypothesis H_0: $\bar{x}=\mu$ (or $\bar{x}_1=\bar{x}_2$) and the size of the difference $|\bar{x}-\mu|$ (or $|\bar{x}_1-\bar{x}_2|$). When the power function is plotted against the difference, the *power curve* of the test is obtained.

The calculation is done before the experiment is carried out, so that the design can be changed if it is insufficiently powerful. If we do the experiment first, and judge the result not significant, we must ask whether the experiment had a good chance of discovering a significant difference if there was one to be found. The power function gives the answer to that question.

The most powerful statistical tests (the Normal, t- and F-tests) give the highest probability of rejection of H_0 for small difference values. In other words, they give us the highest probability of finding a significant difference if there is one. Hence the reason why the tests are very widely used.

The t-test has a slightly lower power curve than that of the Normal test under given circumstances: it is said to have lower *power efficiency*. Hence, we are to some extent less likely to reject a false H_0 by using the t-test unless we use more samples. If test A requires n_a samples and test B requires n_b samples to achieve the same power, the power efficiency of B is $100n_a/n_b$% that of A.

In Section 3.2, we decided how many samples per batch would be required in an analytical control scheme, dependent on the chosen values of α and β. Having done that, we derived a criterion mean response $\bar{x}^*$ which must be exceeded for acceptance of the batch. The power curve looks at this problem in a more detailed way. We will want to be highly confident that we send out product which meets our criterion, for this will doubtless be part of our specification. We shall see that, with few samples per batch, such high confidence can only be achieved if the population value μ is perhaps considerably above $\bar{x}^*$ when our experimental mean $\bar{x}=\bar{x}^*$. As the number of samples analysed increases, this difference $|\bar{x}^*-\mu|$ becomes smaller. More analysis means we need not 'give away' so much material.

Correspondingly, if n is small, there may be an unacceptable level of probability that $\bar{x}=\bar{x}^*$ in fact represents inferior material, with μ considerably below $\bar{x}^*$. As n increases, this probability declines and our customer is less likely to consider our product sub-standard.

3.3.1 Derivation of Power Curves and Sampling Schemes, and the Need for Precision in Manufacture

Assume that a future control procedure will perform n tests on each batch of product, and that a batch will be accepted if the mean result $\bar{x}=\bar{x}^*=95.0\%$ purity. The power curve for the Normal test is given by calculating U values related to $\bar{x}^*$:

$$U=(\bar{x}^*-\mu)\sqrt{n}/\sigma \qquad (2.4b)$$

for fixed $\bar{x}^*$, n, and σ and various values of μ. The Normal distribution table (Appendix 2) gives the probability p that U will be equal to or more than a given value. If U is positive, the probability of accepting the batch is less than 0.5: p is the entry in the table. If $\mu>\bar{x}^*$, U is negative, and the table entry equals $1-p$.

For instance, if the acceptance criterion is that the mean percentage purity over four samples shall be not less than 95.0%, then the p-values obtained via equation (2.4b) for values of population mean μ from 93.0 to 97.0% are shown in Table 3.1. (It is assumed that $\sigma=$unity.)

If we plot p and μ on Normal probability paper, a straight line ensues. The

Table 3.1 *Normal power curve: probability of accepting a batch with mean quality μ for $\bar{x}^*=95.0$, $n=4$, $\sigma=1$*

μ, %	U	P
93.0	4	$\rightarrow 0$
93.5	3	~ 0.001
94.0	2	~ 0.02
94.5	1	0.16
95.0	0	0.50
95.5	-1	0.84
96.0	-2	~ 0.98
96.5	-3	~ 0.999
97.0	-4	$\rightarrow 1$

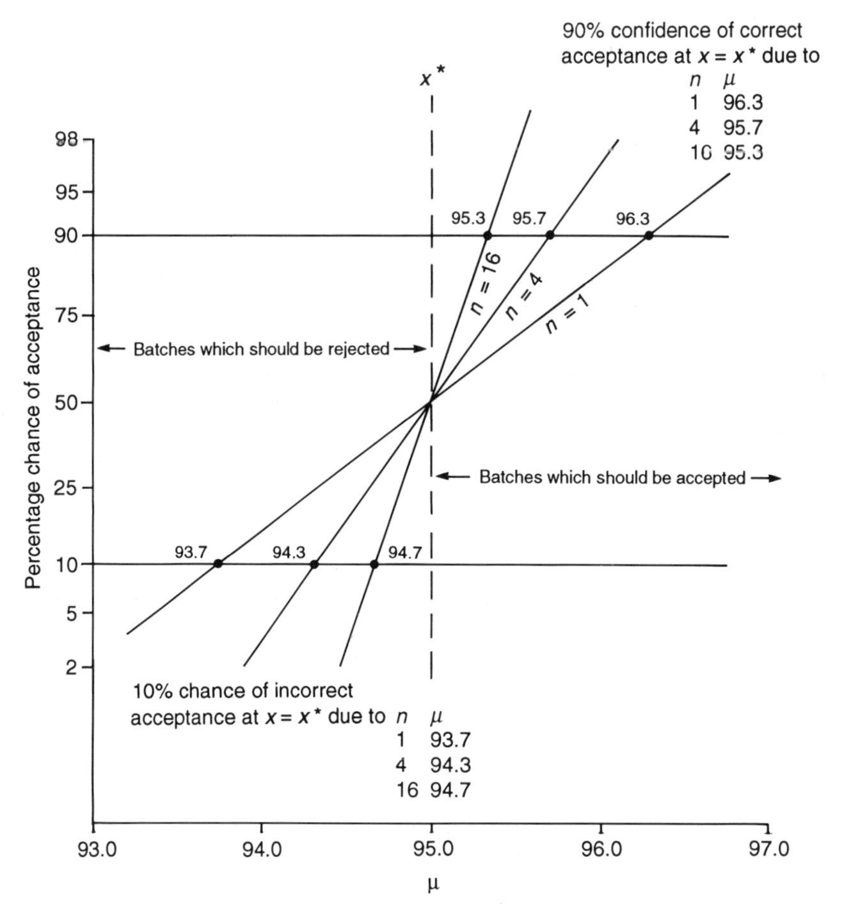

Figure 3.2 *Power curves for single-sided Normal test;* $\bar{x}^*=95.0$, $\sigma=1.0$

steeper the gradient, the greater is the power of the significance test and test circumstances chosen. Figure 3.2 shows the increase of power for the single-sided Normal test with increase of sampling, $n=1$, 4, 16. The necessary μ for 90% probability of acceptance of a batch decreases from 96.3 to 95.7 to 95.3% as n increases $1 \rightarrow 4 \rightarrow 16$. The more samples we are able to analyse, the less we need to produce material much above our criterion. The curve also shows by how much we must improve our purity (for a given sampling scheme) if we wish to increase our probability of accepting a good batch. This corresponds to reducing α, which here is the *producer's risk* of rejecting a good batch.

The 10% chance of incorrectly accepting a bad batch, having recorded $\bar{x}=\bar{x}^*$, corresponds to $\mu=93.7\%$ if $n=1$. As n rises to 4 and 16, μ rises to 94.2 and 94.7%. There is still the same chance of error, but the product is less inferior. This 10% chance equals $\beta=0.10$, and is the *consumer's risk*. To reduce this risk for given μ, we can increase n. These examples illustrate the importance of the producer and consumer agreeing not only a specification but also a sampling scheme.

Once again, because $U \propto \sqrt{n}/\sigma$, halving the standard deviation (of the whole process, not just the analytical work) has the same effect as increasing the number of samples four-fold. The need for maximum precision in manufacture (as well as in analysis) is obvious. There is no point in allowing variability in the process, and trying to compensate by multiple analysis by automated methods. That will increase the cost of testing, the likelihood of tension between the analytical and production departments, and the need for recycling or mixing-off of batches. These last points are self-evident to any one with factory experience; statistics rams them home quantitatively.

Families of power curves (otherwise known as *operating characteristics* curves) have been calculated for Normal and t-tests, to allow for the construction of sampling schemes based on various α- and β-risks of producing material unnecessarily above or unacceptably below $\bar{x}^*$.[8–10] The power curve for a two-tailed test is calculated by using two criteria, $\bar{x}^*_-$ and $\bar{x}^*_+$, below and above which $\bar{x}$ should not fall. The curve, on Normal probability paper, is then a V shape, from a positive slope for $\mu > \bar{x}^*_+$ and a negative slope for $\mu < \bar{x}^*_-$. So long as $\bar{x}$ *lies between* $\bar{x}^*_-$ and $\bar{x}^*_+$, we conclude that the batch μ is not significantly different from the value where the lines cross.[11] Further details on sampling and specifications can be found in more specialized texts.[12]

REFERENCES

1. C. Chatfield, 'Statistics for Technology', 3rd revised Edn., Chapman and Hall, London, 1989, p. 165.
2. G.E.P. Box, W.G. Hunter, and J.S. Hunter, 'Statistics for Experimenters', Wiley, New York, 1979, p. 109.
3. 'Statistical Methods in Research and Production', ed. O.L. Davies and P.L. Goldsmith, 4th revised Edn., Longman, London, 1976, p. 115.
4. B.E. Cooper, 'Statistics for Experimentalists', Pergamon, Oxford, 1969, pp. 73–76, 94–96.
5. 'The Design and Analysis of Industrial Experiments', ed. O.L. Davies, 4th Edn., Longman, London, 1978, pp. 606–611.
6. Reference 3, pp. 94–120.
7. Reference 5, p. 62.
8. C.L. Ferris, F.E. Grubbs, and C.L. Weaver, *Ann. Math. Stat.*, 1946, **17**, 178–197.
9. A.H. Bowker and G.J. Lieberman, 'Engineering Statistics', Prentice-Hall, Englewood Cliffs, NJ, 1959.
10. J.B. Kennedy and A.M. Neville, 'Basic Statistical Methods for Engineers and Scientists', 2nd Edn., Harper and Row, New York, 1976, pp. 389–407.
11. Reference 3, pp. 111–114.
12. See *e.g.* Reference 3, Chapters 5 and 11 and references therein.

CHAPTER 4

Range-finding Experiments

4.1 SORTING A FEW VARIABLES FROM MANY

Many variables may affect the outcome of a chemical analysis or synthesis. Examples include time; temperature; nature, concentration, and order of addition of reagents; nature and concentration of catalysts (and catalyst poisons); nature and concentration of solvents; pH (a specific example of concentration); pressure (another concentration effect); stirring rate; source of materials; nature and intensity of ambient light; ambient atmosphere; nature of interfaces in heterogeneous reactions; sampling method; separation method; operator performance; reactor design. Some of these naturally subdivide; for example, the nature of a solvent may involve viscosity, dielectric constant, surface tension, co-ordinating power, acid–base behaviour. The chemist's knowledge and intuition may exclude many of these from consideration in a particular case. Nonetheless, a considerable number may remain, too many for a full factorial. A preliminary sorting-out is then necessary, and this may be done by Daniel's method of *normal probability plotting*.[1,2] This applies to two-level factorial experiments, and is illustrated by Example 4.1.

Example 4.1

Following a method put forward by Plackett and Burman in 1946, we may represent the eight treatment combinations of a particular fractional factorial design by a matrix in which a positive sign indicates the high level and a negative sign the low level of any given factor: see Table 4.1.

A set of n related experiments can estimate the effects of $(n-1)$ factors or interactions. Here, eight treatment combinations are used to estimate the effects of seven main factors. This is known as a *saturated design*: we devote all our analysis to estimating main factors only. A full factorial for seven factors would require $2^7 = 128$ treatment combinations. Therefore, eight TCs is only one-sixteenth of the full design, and it follows that each effect, ascribed to a factor, can theoretically have up to fifteen other explanations: each factor has fifteen *aliases*. Many of these aliases, but maybe not all, will be zero or insignificant, or anyway unintelligible, and we shall deal with such problems in Sections 6.6, 6.8, and 10.3.2. In a saturated design, to select the (probably) few significant variables from many, we assume the aliases are insignificant.

We then estimate the effect of any factor as the mean of the results at its high level minus the mean of the results at its low level. Hence, if the results x_1 to x_8 are

Table 4.1 *A fractional factorial design for seven variables at two levels*

Experiment number	*Variables* A	B	C	D	E	F	G
1	+	+	+	−	+	−	−
2	−	+	+	+	−	+	−
3	−	−	+	+	+	−	+
4	+	−	−	+	+	+	−
5	−	+	−	−	+	+	+
6	+	−	+	−	−	+	+
7	+	+	−	+	−	−	+
8	−	−	−	−	−	−	−

obtained from experiments 1 to 8 respectively, the effect of changing A, eff_A, is estimated by

$$\frac{x_1+x_4+x_6+x_7}{4}-\frac{x_2+x_3+x_5+x_8}{4}$$

An experiment intended to improve the percentage yield of a product gave results as in Table 4.2.

As a consequence, the estimates of effects of the seven variables are as follows:

Factor	*A*	*B*	*C*	*D*	*E*	*F*	*G*
Effect	+0.58	+6.53	−2.23	+3.28	+13.93	+2.23	+1.23

These effects are plotted in order of increasing magnitude, from most negative to most positive, along the horizontal axis of normal probability paper and then referred to the vertical axis. The horizontal axis has a linear scale while the vertical axis has the normal probability scale (Figure 4.1).

Effects which are not real but are due to random (*i.e.* experimental) error, will plot essentially on a straight line. Those effects which fall off it may be real, and deserve further attention. How far off is sufficient to attract attention can be done by chemical common sense inspection of the effect magnitudes, or (better) by calculating the *coefficient of determination*, the fraction of the total sum of squared deviations of the results accounted for by the effects chosen as significant, or by a

Table 4.2 *Results from a saturated design*

Experiment	*Yield (% theoretical)*
1	90.2
2	81.2
3	87.6
4	91.4
5	95.3
6	73.2
7	83.0
8	71.4

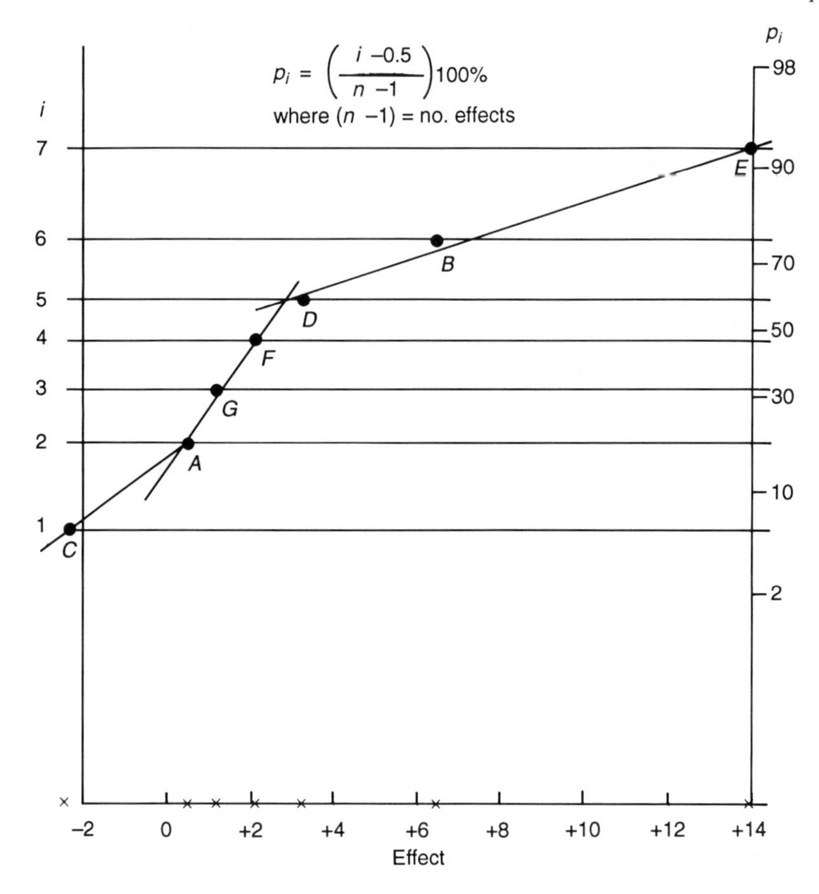

Figure 4.1 *Plotting of effects on Normal probability paper (see also Figure 5.2)*

Daniel *plot of residuals* (the differences between the observed results and those calculated assuming unselected effects were truly zero.[3] The latter procedure is explained in Section 5.7.

The horizontal 'step-lines', onto which successive effects are referred, represent fractional probabilities such that the ith magnitude of effect falls on the p_ith line, determined by $p_i = (i - 0.5)/(n - 1)$, $n - 1$ being the total number of effects.

Figure 4.1 indicates that B, C, and particularly E are significant factors. They should be investigated further, together with their interactions, by carrying out a 2^3 factorial. It is also common for a strongly significant factor such as E to have measurable interactions with other factors, and such interactions as might be predicted to have significance should also be investigated.

Exercise 4.1

Eight treatment combinations were used to examine five factors A, B, C, D, G which might affect a dimerization experiment. The aim was to maximize the ratio of dimer to oligomers, and the results from the eight runs led to the estimates of the effects of the factors shown in Table 4.3.

Plot these effects on normal probability paper, and decide which factors should be investigated further to maximize the ratio.

Note: In Example 4.1 and Exercise 4.1, the reader may judge the answers fairly self-evident. There is no need to stop at this point, however. A method of scrutinizing provisionally identified significant effects, to see how far they account for the experimental data, is described in Section 5.7, and Exercise 4.1 is further discussed in Exercise 6.9.

Table 4.3 *Factor effects in dimerization*

Factor	*A*	*B*	*C*	*D*	*E*	*F*	*G*
Effect	−1.04	0.19	−0.83	−2.70	−0.05	0.39	0.35

E, *F* are so-called *dummy factors* – *i.e.* lines of the Yates analysis which correspond only to interactions.

4.2 A FIRST APPROACH TO THE OPTIMUM

Having identified important variables by the procedure in Section 4.1, the experimenter may still be uncertain, having little knowledge of where the optimum conditions might be. In such a case, we can apply the so-called *Monte Carlo method* of random experiments, making another use of normal probability paper. Such experiments allow a preliminary examination at quite a lot of levels within overall boundaries which are safe and sensible. Consider the two-variable illustration[4] in Example 4.2.

Example 4.2

In an experimental synthesis, thirteen treatment combinations were done, each combination being fixed by choosing *at random* (and separately from one another) one temperature and one time from the following: $T\,^{\circ}\mathrm{C}=26,28,\ldots 54$ and $t\,\mathrm{min}=55,60\ldots 85$. (After any temperature and time are drawn, they *remain available* for further random choice; every combination must have an equal probability of being drawn.) The results, ranked in order, are plotted on the linear axis of the paper against the function $rank/(n+1)$, where n = total number of results. If we write $R=100[rank/(n+1)]$, then $100-R$ is the percentage probability of a yield greater than that obtained at any particular rank. Hendrix's results plot as Figure 4.2, with the straight line drawn by eye.

The best-yield conditions can be the start of an optimization study. In addition, the straight line estimates that there is only about 10% chance of a yield above 70% in the 'experimental space' sampled. If such a yield is unacceptably low, other variables should be looked for which will increase the yield (or one's expectations have to be reduced). It is also possible to estimate the mean and standard deviation of the results by reading off the vertical axis at points corresponding to $R=50$ and 84% (the justification for this comes from the properties of the Normal distribution).

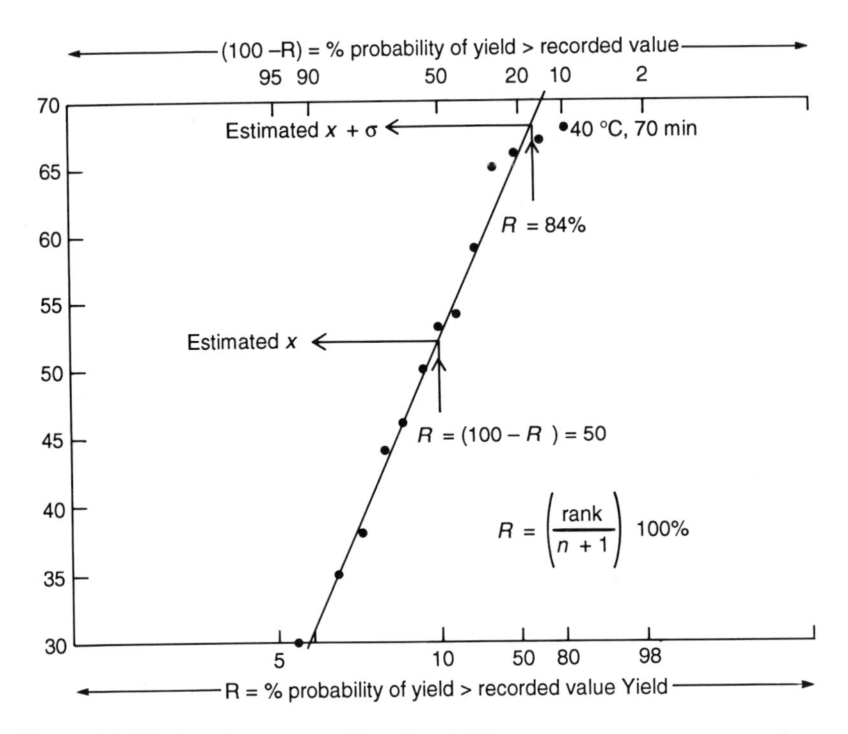

Figure 4.2 *Plotting of responses on Normal probability paper*

Exercise 4.2

Any sufficiently large sampling of a given experimental space should give approximately the same results. Verify the validity of this as follows. Write down the numbers $T = 26, 28, \ldots 54$ on slips of paper, fold them and place them in a beaker. Do similarly with $t = 55, 60, \ldots 85$. Draw one slip from each beaker, record the treatment combination, and put the papers back. In this way, record thirteen such combinations. Then read off the yield for these experiments from the response surface which represents the total behaviour of the system (Figure 4.3). (The points shown thereon are Hendrix's results.) Rank the results, plot them as in Figure 4.2 and identify your best-yield conditions, probability of a yield above 70%, $\bar{x}$, and σ. They should be close or fairly close to the values on Figure 4.2.

There is little point in running more than twenty random trials. If F = fraction of chosen experimental space which will prove useful, then $1 - F$ = fraction which is not useful. Hence, $(1-F)^n$ is the probability of all of n trials not being useful. It follows that $1-(1-F)^n$ is the probability that at least one random trial lands in useful space. Let us assume that we are so unsure of the useful space that our best guesses of the levels to use lead to the useful space being only 10% of the total tested ($F = 0.10$). We would then have $1-(1-F)^n$ changing from 0.65 to 0.79 to 0.88 as n changes from 10 to 15 to 20. There is 88% certainty that at least one of the twenty trials will land in useful space. If our guesses are better, so that $F = 0.20, 1-(1-F)^n$

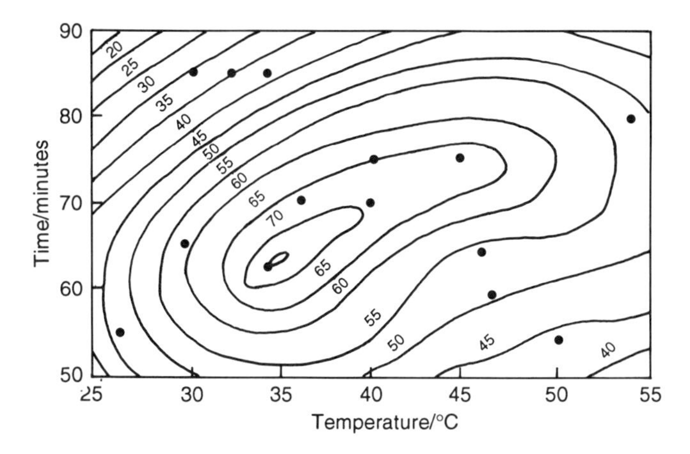

Figure 4.3 *Response surface of a two-variable system (Exercise 4.2)*

varies from 0.67 to 0.89 to 0.96 as n varies from 5 to 10 to 15. We are virtually assured of at least one useful result from fifteen trials.

It is obviously important to have as good an initial idea as possible. This is not only because we then need to do less work. In addition to this, it is not easy to ensure that trials are carried out as intended, and the more numerous the variables and the more numerous the trials, the less likely is it that things will go according to plan. There is also a very important reason for trying to exclude non-significant variables from our investigations. If, in the words of Box and his colleagues,[5] 'we are to use statistics to catalyse the creativity of engineers and scientists, they must know what factors to reason about'.

REFERENCES

1. C. Daniel, *Technometrics*, 1959, **1**, 311–341.
2. G.E.P. Box, W.G. Hunter, and J.S. Hunter, 'Statistics for Experimenters', Wiley, New York, 1979, pp. 329ff.
3. C. Daniel, 'Applications of Statistics to Industrial Experimentation', Wiley, New York, 1976, pp. 73ff.
4. C. Hendrix, *Chemtech*, August, 1980, 448–497.
5. G.E.P. Box, S. Bisgaard, and C. Fung, *Qual. Reliability Eng. Int.*, 1988, **4**(2), 124–131.

CHAPTER 5

Factorial Designs at Two Levels

The factorial design of experiments has been described[1] as the most powerful technique in technological research. The simplest designs, using factors at two levels, are of great practical importance, and were introduced in Section 1.3 of this text. We saw there that for a given amount of work, compared with the classical approach, they give more estimates of the effect of a factor (and hence we have greater precision in our knowledge of that effect) and additionally they tell us of the effects when factors interact with one another. There is also another advantage that they extend the range of validity of our conclusions. For example, the logic and economy of factorial experiments may allow us to use several batches of raw material in an investigation. This is *not* a disadvantage; it provides a wider inductive basis to reach conclusions on, and we can apply those conclusions with much more confidence than results based on just one batch.

5.1 COMPLETION OF YATES TABLE ANALYSIS OF VARIANCE

The estimation of the effects of the factors and interactions was described in Sections 1.3.1–1.3.3. We can now complete the Yates table analysis of a 2^3 experiment begun in Table 1.2: see Table 5.1.

The general expression for the sum of squares SS_x for any effect eff_x in a 2^n factorial experiment with r replications is given by

$$SS_x = \frac{(r2^{n-1} eff_x)^2}{r2^n} = \frac{f_x^2}{N} \qquad (5.1)$$

where f_x is the number in the final column of analysis corresponding to x and N is the *total* number of treatment combinations. Hence, in a duplicated 2^3 experiment, $\underline{3} = 8eff_x$ and $SS_x = (2.4eff_x)^2/2.8 = 4eff_x^2$. In a triplicated 2^4 experiment, $\underline{4} = 24eff_x$ and $SS_x = 12eff_x^2$.

5.1.1 Assessment of Significance of Effects by Comparison of Mean Squares (the *F*-test)

A *sum of squares* is the sum of the squares of the deviations of a set of experimental results about their mean. When we divide a sum of squares by the appropriate degrees of freedom, ϕ, we obtain the corresponding variance or *mean square* s^2 (see equation 2.5). To decide whether one effect is significantly bigger than another, we calculate the ratio of their mean squares, with the larger mean square as the

Table 5.1 *Completion of Yates' table analysis for an unreplicated 2^3 experiment (see Table 1.2)*

Column of analysis $\underline{3}$	*Effect* $=\underline{3}/4$	*Sum of squares* $=\underline{3}^2/8$	*Mean square = sum of squares/ degrees of freedom*
(Total)			
$4eff_A$	eff_A	$2eff_A^2$	
$4eff_B$	eff_B	$2eff_B^2$	as
$4eff_{AB}$	eff_{AB}	$2eff_{AB}^2$	
$4eff_C$	eff_C	$2eff_C^2$	sum of squares
$4eff_{AC}$	eff_{AC}	$2eff_{AC}^2$	
$4eff_{BC}$	eff_{BC}	$2eff_{BC}^2$	(see Section 5.3)
$4eff_{ABC}$	eff_{ABC}	$2eff_{ABC}^2$	

numerator. The result is an experimental F-value (if in doubt, refer back to Section 2.6.5). This value can be compared with F-table values, which can be expected to occur merely by chance. If our experimental F-value is greater than the value expected in (usually) 95% of cases, we say that one effect is 'significantly' greater than the other. If the experimental value would be exceeded by chance only 1% of the time, we say there is a 'highly significant' difference.

The Yates table is a particularly convenient analysis of variance, to divide up the total variance (*total mean square*) of a factorial experiment into its several sources. If the experiment is not replicated, *all* the variance is subdivided between the effects. However, if it is replicated, *some* of the total variance is ascribed to each effect, and the remainder is ascribed to experimental error. Each sum of squares is divided by the appropriate value of ϕ. The numerator in any experimental F is an effect mean square. The remainder sum of squares is divided by its value of ϕ to give the *error mean square* (otherwise known as the *error*, or *residual*, *variance*) and this is the denominator of F.

5.1.2 Degrees of Freedom and Calculation of Mean Squares

A 2^3 experiment has (2^3-1) degrees of freedom, and Yates analysis divides the total variation in the results into the seven effects. It follows that each effect has one degree of freedom; hence, for any effect, the mean square equals the sum of squares. If the experiment is duplicated, however, the sixteen results have fifteen degrees of freedom. But the seven effects still take only one each, and therefore eight degrees of freedom are assigned to the error. In a duplicated 2^3, therefore, the error mean square is the remainder sum of squares divided by eight. In a triplicated 2^3 (see Example 5.1), the remainder sum of squares has sixteen degrees of freedom.

5.1.3 Calculation of the Error Sum of Squares

We calculate the remainder sum of squares by deducting the sum of the sums of squares for all the effects from the total sum of squares for all the experimental results about their mean, *TSS*. If all the results $x_1, x_2 \ldots x_p$ give a mean $\bar{x}$, the total sum of squares is given by

$$TSS = \sum_{i=1}^{p} (x_i - \bar{x})^2 \tag{5.2}$$

or, more conveniently

$$TSS = \left\{ \sum_{i=1}^{p} x_i^2 - \frac{[\sum_{i=1}^{p} x_i]^2}{p} \right\} \tag{5.3}$$

That is to say, we calculate *TSS* by

(i) adding together the squares of all the results, and

(ii) subtracting from the result in (i), the value of (the square of the total of all the results, divided by the number of results).

The sum of the effects, ΣSS, is obtained simply by adding all the sums of squares from the Yates table. Then,

$$\text{remainder sum of squares} = TSS - \Sigma SS \tag{5.4}$$

As said before, the remainder divided by the appropriate ϕ is the error variance, which represents all unidentified sources of variation and which we use to test whether a given effect is significant. The error variance is therefore our baseline of comparison. If our experimental work is erratic, and/or there is an unidentified source of variation, the error variance may become large and we may fail to detect significant effects.

5.2 THE DESIGN AND ANALYSIS OF A 2^3 FACTORIAL EXPERIMENT IN TRIPLICATE

Example 5.1 is offered as a detailed model and, for completeness, includes some important points made elsewhere in this text.

Example 5.1

A titrimetric investigation[2] of the redox reaction between ascorbic acid and ferric iron was set up, to assess (i) whether the determination of ferric iron by this method is precisely stoichiometric, and (ii) whether the reaction might alternatively be used for the determination of the vitamin C.

These two versions of the operation (Figures 5.1a and b) constitute different orders of addition of the reagents, and such changes commonly cause changes in chemical outcomes. In a titration or other case where one reagent is added slowly to another, we can differentiate crudely between 'adding a little A to a lot of B' and 'adding a little B to a lot of A.' If the reaction is essentially quantitative and is sufficiently rapid, this little/lot relationship exists until very near the end-point. Even if it is not, the reagent initially in the flask must be in excess at least until half the stoichiometric equivalent of the second reagent has been added. Effects due to

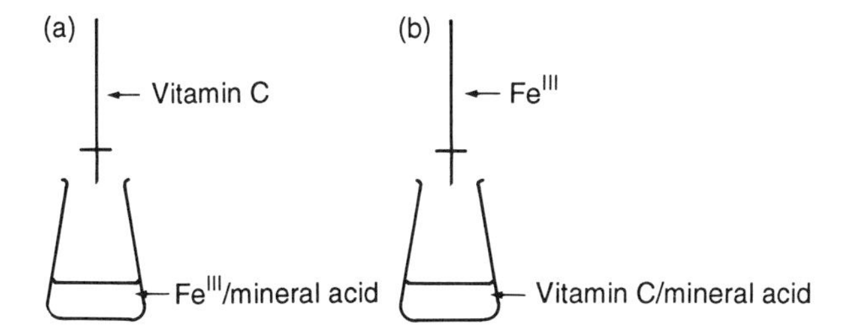

Figure 5.1 *Two versions of the iron*(III)*–ascorbic acid titration*

reagent excess might therefore differ from one version of the reaction to the other.

Based on the existing literature, three variables (temperature, acidity, and reagent concentration) were examined, each at two levels (Table 5.2). The set of eight treatment combinations was carried out three times in different *random* orders; Table 5.3 shows them for vitamin C in the flask.

The responses for any one treatment combination were summed, and the whole then tabulated in the *fixed* order required by the Yates table. For instance, for the titration of ferric iron by vitamin C:

Treatment combination	*Response* (cm^3)
(1)	75.22
a	74.13
b	75.52
ab	74.73
c	75.29
ac	74.79
bc	75.62
abc	74.99

The Yates table analyses and analyses of variance were as shown in Tables 5.4 and 5.5.

Table 5.2 *Levels of variables chosen for the vitamin C/Fe^{III} reaction*

	Variable A (*temperature* °*C*)	*Variable B* (*mineral acidity initially in flask*) $M H^+$	*Variable C* (*reactant concentration initially in flask*) *N*
(*a*, *b*, *c*) upper level	80	0.8	0.15
[(1), (1), (1)] lower level	40	0.1	0.10

Table 5.3 *Random experimental orders*

Run	*Run order*
1	(1), *ab abc a c b ac bc*
2	*bc ab ac abc b c* (1) *a*
3	*b c* (1) *abc ac bc a ab*

Table 5.4 *Analysis of titration volumes for titration with Fe^{III} in flask*

TC	*Sum of three titres*	*Columns of analysis* 1	2	3	*Effect* $=\underline{3}/4\times 3$	*Sum of squares* $=\underline{3}^2/3\times 8$
(1)	75.22	1.35	3.60	8.29		
a	74.13	2.25	4.69	−3.01	−0.2508	0.3775
b	75.52	2.08	−1.88	1.43	0.1191	0.0852
ab	74.73	2.61	−1.13	0.17	0.0141	0.0012
c	75.29	−1.09	0.90	1.09	0.0908	0.0495
ac	74.79	−0.79	0.53	0.75	0.0625	0.0234
bc	75.62	−0.50	0.30	−0.37	0.0308	0.0057
abc	74.99	−0.63	−0.13	−0.43	0.0358	0.0077

Code: deduct 74.00

Mean titre value indicated 25.00 cm^3 0.1 N vitamin C = 24.99 cm^3 0.1 N Fe^{III} – contrast Table 5.5.

Effect	*Sum of squares*	*Mean square*	$\left(\frac{\textit{Mean square}}{\textit{Residual variance}}\right)$ = *F experimental*
Factor *A*	0.3775 ($\phi=1$)	0.3775	10.97 {significant at 1% level (one-tailed test)}
Factor *B*	0.0852 ($\phi=1$)	0.0852	2.48
Factor *C*	0.0495 ($\phi=1$)	0.0495	1.44
Two-fold interaction *AB*	0.0012 ($\phi=1$)	0.0012	0.03
Two-fold interaction *AC*	0.0234 ($\phi=1$)	0.0234	0.68
Two-fold interaction *BC*	0.0057 ($\phi=1$)	0.0057	0.17
Three-fold interaction *ABC*	0.0077 ($\phi=1$)	0.0077	0.22
Sum of effects	0.5502 ($\phi=7$)		
Remainder	0.5511 ($\phi=16$)	0.5511/16 = 0.0344 = error variance	
Total	1.1013 ($\phi=23$)		

Note on coding: since every estimate of an effect is a difference of two values, we can reduce the figures by deducting a fixed amount from each value. For instance, above, 74.13 − 75.22 = 0.13 − 1.22 = −1.09 = *a* − (1).

Table 5.5 *Analysis of titration volumes for titration with vitamin C in flask*

TC	*Sum of three titres*	*Columns of analysis* 1	2	3	*Effect* $=\underline{3}/4\times 3$	*Sum of squares* $=\underline{3}^2/3\times 8$
(1)	74.27	0.79	0.83	2.08		
a	75.52	0.04	1.25	0.42	0.035	0.0074
b	73.97	0.90	0.35	−1.30	−0.108	0.0704
ab	74.07	0.35	0.07	−0.08	−0.0067	0.0003
c	74.45	0.25	−0.75	0.42	0.035	0.0074
ac	74.45	0.10	−0.55	−0.28	−0.0233	0.0033
bc	74.14	0.00	−0.15	0.20	0.0167	0.0017
abc	74.21	0.07	0.07	0.22	0.0183	0.0020

Code: deduct 74.00

Mean titre value indicated 2.500 cm^3 0.1 N vitamin C = 24.76 cm^3 0.1 N Fe^{III} – contrast Table 5.4.

Effect	*Sum of squares*	*Mean square*	$\left(\frac{\textit{Mean square}}{\textit{Residual variance}}\right)$ $= F$ *experimental*
Factor *A*	0.0074	0.0074	1.06
Factor *B*	0.0704	0.0704	10.06 {significant at 1% level (one-tailed test)}
Factor *C*	0.0074	0.0074	1.06
Two-fold interaction *AB*	0.0003	0.0003	0.04
Two-fold interaction *AC*	0.0033	0.0033	0.47
Two-fold interaction *BC*	0.0017	0.0017	0.24
Three-fold interaction *ABC*	0.0020	0.0020	0.29
Sum of effects	0.0925		
Remainder	0.1112	0.0070 = error variance	
Total	0.2037		

5.3 ASSESSING THE SIGNIFICANCE OF EFFECTS IN REPLICATED EXPERIMENTS

The analysis of variance has given us the F-values for the various effects (right-hand column, Tables 5.4 and 5.5, lower section). In this particular experiment, they have degrees of freedom $\phi = 1$ for each effect mean square – the effect variances – and 16 for the residual variance (residual mean square).

We now compare the experimental values with $F_{\varphi} = 1,16$ in the F-table. This lists F-values which can be expected to occur merely by chance, without any systematic effects being present. The table is always set out with ϕ corresponding to the greater mean square shown horizontally and that for the lesser variance

Table 5.6 *Probability points of the F-distribution for a one-tailed test*

% probability	ϕ_d	ϕ_n 1	5	10	15	20
10		39.9	57.2	60.2	61.2	61.7
5	1	161	230	242	246	248
2.5		648	922	969	985	993
1		4052	5672	6056	6157	6209
10		4.06	3.45	3.30	3.24	3.21
5	5	6.61	5.05	4.74	4.62	4.56
2.5		10.0	7.15	6.62	6.43	6.33
1		16.3	11.0	10.1	9.72	9.55
10		3.28	2.52	2.32	2.24	2.20
5	10	4.96	3.33	2.98	2.84	2.77
2.5		6.94	4.24	3.72	3.52	3.42
1		10.0	5.64	4.85	4.56	4.41
10		3.07	2.27	2.06	1.97	1.92
5	15	4.54	2.90	2.54	2.40	2.33
2.5		8.20	3.58	3.06	2.86	2.76
1		8.68	4.56	3.80	3.52	3.37
10		2.97	2.16	1.94	1.84	1.79
5	20	4.35	2.71	2.35	2.20	2.12
2.5		5.87	3.29	2.77	2.57	2.46
1		8.10	4.10	3.37	3.09	2.94

ϕ_n = degrees of freedom, numerator
ϕ_d = degrees of freedom, denominator

vertically. In the body of the table are 'chance-occurrence' values of F on horizontal lines corresponding to different probabilities of purely chance occurrence. An abstract from the table is reproduced as Table 5.6. A detailed table is given in Appendix 4.

In Tables 5.4 and 5.5, each figure in the Effect column is the mean of twelve estimates (because the 2^3 factorial was triplicated). Therefore, although the significant effect in titrating vitamin C was only $-0.10\ cm^3$ (*i.e.* about 0.4% of the mean titre), considerable confidence can be placed in it as a systematic effect of changing the acidity. If a variation of $0.10\ cm^3$ is observed in the normal two or three titrations, it may be ignored or averaged as experimental error. But not here: we have obtained as many as twelve 'opinions' which, averaged out and compared with the evaluated experimental error, give it significance.

Exercise 5.1

A triplicated 2^3 experiment gave the following results. Analyse by Yates' method and give the significance level of each effect by using the F-test.

(1)	53.4	47.4	43.2	*c*	50.2	50.2	43.6
a	31.2	25.2	33.6	*ac*	22.0	22.0	22.0
b	23.2	17.2	19.6	*bc*	26.0	26.0	20.0
ab	16.2	10.2	15.6	*abc*	33.8	33.8	34.4

giving grand total 720.0, mean 30.0, and treatment totals (1) 144.0, *a* 90.0, *b* 60.0, *ab* 42.0, *c* 144.0, *ac* 66.0, *bc* 72.0, *abc* 102.0.
(Hints: Use the treatment totals in the Yates table. If you wish to code these totals, note that the mean result is 30.0, and the mean total is therefore 90.0. Code by deducting this amount. If you do, the figure in the top row of the final column of analysis must be zero.)

Answer

The effect magnitudes and sums of square are

	A	*B*	*AB*	*C*	*AC*	*BC*	*ABC*
Effect	−10	−14	+12	+4	+2	+8	+6
Sum of squares	600	1176	864	96	24	384	216 ($\phi=1$)

$$\text{The total sum of squares} = \left\{ \sum_{1}^{24} x_i^2 - \frac{\left[\sum_{1}^{24} x_i\right]^2}{24} \right\}$$

$$= 3543.36 \ (\phi = 23)$$

Total treatment sum of squares = 600 + 1176 + ... 216 = 3360 ($\phi=7$). Therefore remainder sum of squares = 3543.36 − 3360 = 183.36 ($\phi=16$). Hence residual (error) variance = 183.36/16 = 11.46.

Dividing the effects sums of squares (= effects mean squares) by error variance gives experimental $F_{1,16}$ values:

Effect	*A*	*B*	*AB*	*C*	*AC*	*BC*	*ABC*
F_{expt}	52.4	102.6	8.4	75.4	2.1	33.5	18.8
Probability of chance occurrence (%)	$\ll 1$	$\ll 1$	<2.5	$\ll 1$	>10	$\ll 1$	$\ll 1$

AC has no significant effect; *AB* has a significant effect; the other factors and interactions have highly significant effects.
[Note: if the analysis is carried out on uncoded data, the top – *i.e.* (1) – row of the sum of squares column shows the second term of the total sum of squares expression, $\left(\sum_{1}^{N} x_i\right)^2 \Big/ N$, where N is the total number of results. Therefore, if it is convenient to use uncoded data, this term is calculated automatically by the Yates procedure.]

5.4 CHEMICAL INTERPRETATION OF STATISTICALLY DERIVED CONCLUSIONS

Factorial experiments offer the possibility of considerable insight into chemical reactions. For instance, in the ferric iron–ascorbic acid experiments,

(i) With Fe^{III} in the flask, an increase in temperature from 40 to 80 °C was found to be highly significant; it reduced the vitamin C titre for a given amount of iron (and, by observation, it speeded the reaction up).

(ii) With vitamin C in the flask, an increase in initial mineral acidity from 0.1 to 0.8 M H^+ was found highly significant; it decreased the Fe^{III} titre for a given amount of vitamin C (and, by observation, slowed the reaction down). Hence altering the order of addition changed the significant variable from temperature to acidity.

(iii) Mean titre values showed that, with Fe^{III} in the flask, 25.00 cm^3 vitamin C was equivalent to 24.99 cm^3 Fe^{III}. (This was in good agreement with the literature, where stoichiometric equivalence had been claimed for this titration at 50 °C and about 0.5 M H^+.) But with vitamin C in the flask, 25.00 cm^3 of it was equivalent to only 24.76 cm^3 Fe^{III}.

Can we understand this change of significant variable, and can we rationalize why the vitamin C was about 1% less efficient when it was in the flask rather than in the burette?

My explanation is as follows:

Let W = vitamin C; X = Fe^{III}; Y = dehydrovitamin C; Z = some further oxidation product(s) from vitamin C. When Fe^{III} is titrated (at least until near the end-point), a little W + much X → a little Y (+ much X remaining). Y may react further with X to give Z – some dehydrovitamin C may reduce some Fe^{III}. But when vitamin C is titrated, a little X + much W → a little Y, but negligible X is now present: Y remains.

When vitamin C is in the burette, therefore, some of the dehydrovitamin may be effective in reducing Fe^{III} and a given amount of Fe^{III} will be reduced by less vitamin C than when Fe^{III} is the titrant. *This is observed* from the mean values.

If the dehydrovitamin/Fe^{III} reaction is slow at either 40 or 80 °C, the results from the two versions will not differ by much. A mean difference of 1% was observed. But on changing from 40 to 80 °C where the further reaction is possible (Figure 5.1a), the further oxidation of the dehydro-compound may become noticeable. (Further oxidation is of course possible: alkaline permanganate produces oxalate and *l*-threonate.) If so, the vitamin C titre should decrease as temperature increases. *This is observed* (eff_A is negative in Table 5.4). In the other titration (Figure 5.1b) this temperature dependence was not observed. This is explicable by the absence of the slow second stage.

When Fe^{III} was the titrant, increase of acidity decreased the observed titres significantly and slowed the reaction down. This is reconcilable with the well-known fact that vitamin C is much less quickly oxidized in acidic media. Therefore, at an earlier stage in the titration, it will reach the point where Fe^{III} is likely to remain combined with the thiocyanate indicator for such a time that the observer will conclude that the end-point has been reached. The lack of acidity dependence in version 5.1a is explicable by the fact that, since the vitamin C was added in small increments to the flask, its concentration at any stage must have been low, and therefore concentration reduction by increased acidity cannot ever be more than slight.

The factorial design does not prove the validity of these ideas. But, at least, the statistical analysis can intuitively be reconciled with one's general chemical understanding. It is important to try to make chemical sense of the statistical data: simply stating percentage probabilities will leave unchallenged the possibility that the remaining probability of chance occurrence may mean that the effects are not real at all. This is particularly important if the chemistry is new to us; as the statistician Martin Moroney said,[3] 'to a new-born babe the conduct of a Borgia may be the behaviour to be expected of a Pope'.

5.5 SOME GENERAL RULES FOR THE OPERATION OF FACTORIAL EXPERIMENTS

The investigation described in Section 5.2 followed these rules:

1. The individual titrations were done in *random* order (see Section 2.8 if in doubt).
2. The titrations were in fact divided between two workers. If all the hotter flasks had been held by one worker, and all the cooler ones by the other, there would have been confounding between the temperature effect and any difference between operators. Each operator must do four titrations (in any run) including two high *A*s, two high *B*s, and two high *C*s.
3. The Yates analysis was then done in *fixed* order, to make the arithmetic work.

Shorter random run orders can be decided by putting TC symbols on folded bits of paper into a beaker, shaking them up and drawing them out. If the order does not look random, screw them up and shake again. Random number tables or programs are usually used for larger runs.[4,5] To test for acceptable randomness, we divided the sets of eight (see Table 5.3) with a vertical line midway through them. We found we had seven high-A, seven high-B, and six high-C TCs in the front half of the design, compared with the true random $6+6+6$. We decided that this was near enough (laborious repetition of choosing may be necessary to achieve randomness, even with random-number tables). Each block of eight was then divided into two halves obeying the criterion given in point 2 earlier in this Section. The method to do this is explained in Section 6.4.

Statisticians are very anxious about possible bias in results, and rightly so. We, as analysts, were looking for small differences between large numbers (small variations of titre), and some of the experimental conditions proved to give very sluggish end-points, where great patience had to be exercised. This was therefore a situation *par excellence* where one must not prejudge the issues – where, say, today's (1) experiment must not be influenced by remembering yesterday's results. We therefore did the three runs on three consecutive days, using different standard solutions on each day – for which we did not calculate the precise molarities until afterwards. We did not work out the results until all the runs were completed and we did not tell one another of difficulties or ease with individual titrations.

5.6 ASSESSING THE SIGNIFICANCE OF EFFECTS FROM UNREPLICATED EXPERIMENTS

In the examples dealt with above, the effect mean squares were divided by the error variance, which was possible because there was replication of the factorial. If there had been only one run (*i.e.* eight TCs), then the seven effects would have taken up all the seven degrees of freedom of the design. In practice, this means that all the variations in the results have to be assigned to the seven effects; therefore there is no error variance, and we cannot calculate F as we did. (Which is equivalent to the chemist saying he replicates his experiments to check his reproducibility!) Some effect or effects from within the seven must be used as an estimate of the error variance. In a 2^3 experiment, the highest interaction is ABC, and eff_{ABC} ($\phi = 1$) can be taken as the estimate of error. But the F-table shows that only very large experimental values of F can show significance under such a circumstance. Therefore, it is customary to pool a number of interactions to represent error, or to use all the interactions and thereby estimate the factors only.

The choice of high-order interactions for this purpose is because they are often hard to interpret, and in cases where they are assessed they are often non-significant. Occasionally, however, real and meaningful higher-order interactions occur, and two difficulties can arise. Including them in the error variance would reduce the likelihood of finding other significant effects. Also, there is a temptation to calculate all the effects and then be selective in choosing only the smallest interaction effects for the error variance, which may lead to spurious significance. The choice of interactions should not involve hindsight; it should be made before the experiments are done, based on prior knowledge or reasonable assumption that they will not be significant. Both difficulties are avoided by using Daniel's normal probability plotting of effect magnitudes (Section 4.1 above). These points are illustrated in Examples 5.2 and 5.3 and the corresponding Exercises.

Example 5.2: identifying significant effects by analysis of variance

An unreplicated 2^3 experiment gave results as follows:

TC	(1)	*a*	*b*	*ab*	*c*	*ac*	*bc*	*abc*
Response	53	18	46	20	101	44	89	41

Discussion

These yield effects and mean squares for the effects as follows:

	A	*B*	*AB*	*C*	*AC*	*BC*	*ABC*
Effect	−41.5	−5.0	+4.5	+34.5	−11.0	−2.5	0
Mean square	3445	50	41	2381	242	13	0

Various conclusions can be reached, depending on prior assumptions. For example, if MS_{ABC} had been chosen as the error variance, every experimental F would be indefinitely large, and every effect would be significant; whereas if MS_{AC} had been chosen, no effect would be significant. Most combinations of interactions give high significance or significance to A and C, while taking $\frac{1}{3}(MS_{AB} + MS_{BC} + MS_{ABC})$ as the error variance also gives significance to AC. The four-interaction estimate of error variance identities both A and C as highly significant ($\alpha < 0.01$) but of course cannot tell us about AC. We can compare these conclusions with those from Daniel's method (Example 5.3).

Exercise 5.2a

Assign high significance, significance, or possible significance to A and C assuming the following choices for error variance: (i) error variance $=\frac{1}{2}(MS_{AB}+MS_{BC})$, (ii) error variance $=\frac{1}{2}(MS_{AB}+MS_{AC})$. Show why $\alpha<0.01$ was assigned to A and C in the last paragraph.

Answer

(i) $F_A=127$, $F_C=88$ with $\phi=1,2$ (one-tailed), $\alpha_A<0.01$ (highly significant), $\alpha_C<0.025$ (significant)
(ii) $F_A=24.3$, $F_C=16.8$ with $\phi=1,2$ (one-tailed), $\alpha_A<0.05$ (significant), $0.05<\alpha_C<0.10$ (possibly significant)

Four-interaction estimate of error variance $=74$; $F_A=46.6$, $F_C=32.2$ with $\phi=1,4$. F_{table} values are 21.2 ($\alpha=0.01$) and 74.1 ($\alpha=0.001$).

Exercise 5.2b

An unreplicated 2^3 factorial to estimate main factors only gave the following results:

(1) 1, *a* 2, *b* 3, *ab* 4, *c* 4, *ac* 6, *bc* 7, *abc* 9

Take the mean of the interaction mean squares as the error variance and show that eff_A was significant, eff_B highly significant, and eff_C very highly significant.

Answer

Mean squares A 4.5, B 12.5, AB 0, C 32, AC 0.5, BC 0.05, ABC 0.

	$F_A=18$	$F_B=50$	$F_C=128$ ($\phi_1=1$, $\phi_2=4$)
one-tailed	$0.025>\alpha>0.01$	$0.01>\alpha>0.001$	$0.001>\alpha$

Example 5.3: identifying significant effects by Daniel's method

Plotting the effects from Example 5.2 gives Figure 5.2.

Discussion

Drawing a straight line by eye on Figure 5.2 might identify A and C as the only significant effects, or might suggest AC also. Effects which are merely random error are expected to fall approximately on a straight line; the corresponding effect magnitudes are experimental estimates of non-existent systematic effects (*i.e.* of zero). Making a final decision about significant effects can therefore be attempted by a method due to Daniel, by substituting zeros for the experimental effect values assumed to be on the line, and working backwards by *reverse Yates technique* to calculate what the model produces as initial experimental values (see Section 5.7). If there is little difference between the observed and calculated (*fitted*) values, our model largely accounts for the observed behaviour. If the model is a good one, the differences, plotted on normal probability paper, should give an approximately

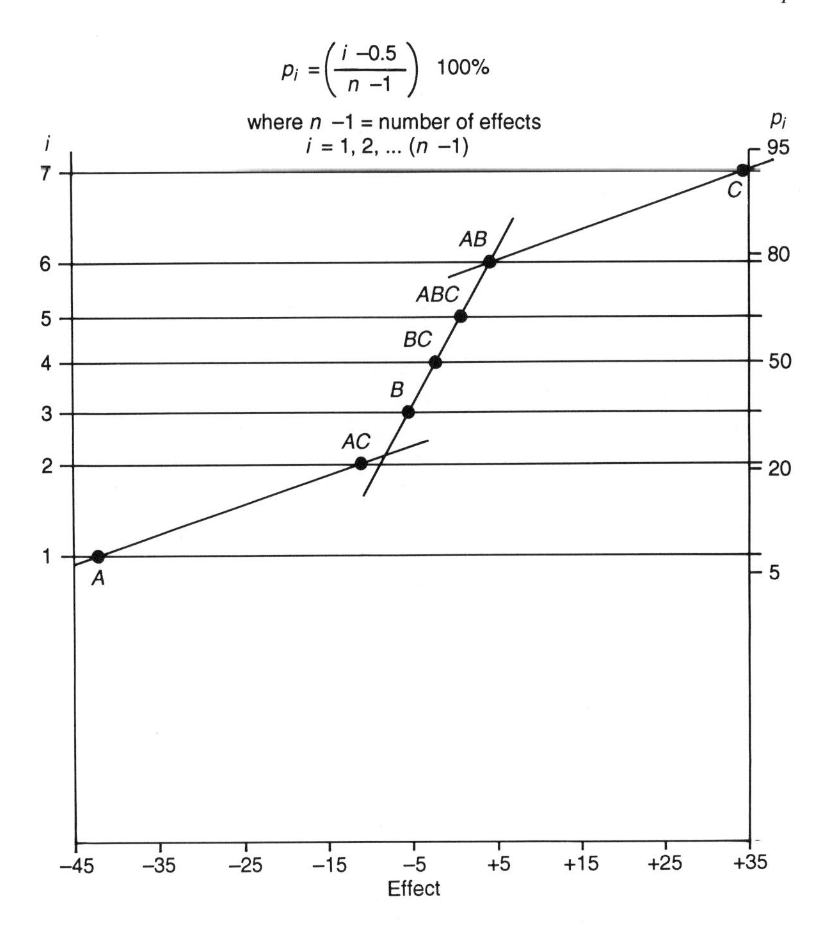

Figure 5.2 *Identification of significant effects in an unreplicated experiment*

straight line, and there should be no visible mathematical relation between individual differences and the experimental results to which they correspond.[6,7]

5.7 REVERSE YATES TECHNIQUE AS A CHECK OF PROCESS MODELS

It can be shown that if the effect values and twice the mean result from an unreplicated factorial design are listed in the reverse of the usual Yates order, and the standard addition and subtraction in pairs is carried out n times for n factors, then the numbers in the nth column of analysis are twice the original experimental responses. [You may care to check this with the data from Example 5.2, writing the effects and twice the mean in the order (top) *ABC* 0, *BC* −2.5, *AC* −11.0, *C* +34.5, *AB* +4.5, *B* −5.0, *A* −41.5, twice mean 103 (bottom) and deriving the responses *abc* 41, *bc* 89, *ac* 44, *c* 101, *ab* 20, *b* 46, *a* 18, (1) 53.]

This procedure is used in model-fitting as illustrated by Example 5.4.

Example 5.4

Assuming from Figure 5.2 that A, C, and AC are significant, the effects and twice the mean from Example 5.2 are listed in reverse order (Table 5.7) with other effects listed as zero.

The differences are now plotted on normal probability paper as shown on Figure 5.3. The non-linear axis is subdivided according to $p_i = 100(i-0.5)/m$, where m is the total number of responses (here, 8). The differences lie close to a straight line, confirming the conjecture that the B, AB, BC, and ABC effects are explicable as random 'noise.' The order of the differences is also noticeably different from that of the associated effects. If this were not the case, the data would need to be transformed in some way.[8–12]

If Y represents the experimental response, a better fit of the residuals may be achieved by *transformation* to, *e.g.*, $\log Y$, $\log(Y-c)$, $\sqrt{Y}$, $1/Y$, or $1/\sqrt{Y}$. Daniel[8] discussed an unreplicated 2^4 experiment in which the model first chosen (B, C, D, BC, CD taken as significant effects) gave several markedly off-line points on the residual graph. In addition, the difference $Y-y$ increased with increasing Y. Eight transformations of Y were looked at, based on Y^a and $\log(Y-c)$, plotting responses to identify probable significant effects and residuals to test model-fitting. It was concluded that $\sqrt{Y}$ with A, B, C, D, BC as significant effects was best, with $\log(Y-1)$ and the same effects very slightly inferior. Neither gave any trend of $Y-y$ with Y, implying that the residuals were truly random 'noise.' At this point, it is necessary to decide whether the square root or the log form is physically more realistic.

Exercise 5.4

Take the effects and twice the mean from Example 5.2, assume that only the main factors A and C are significant, and use the reverse Yates technique and Daniel's plot to decide whether this is a better model than that which also includes AC.

Table 5.7 *Reverse Yates analysis for model fitting*

	Effect	*Columns of analysis* 1	2	3	*Fitted response* $=3/2=y$	*Experimental response* $=Y$	*Difference* $=Y-y$
ABC	0	0	23.5	85.0	42.5	41	−1.5
BC	0	23.5	61.5	190.0	95.0	89	−6.0
AC	−11	0	45.5	85.0	42.5	44	+1.5
C	+34.5	61.5	144.5	190.0	95.0	101	+6.0
AB	0	0	23.5	38.0	19.0	20	+1.0
B	0	45.5	61.5	99.0	49.5	46	−3.5
A	−41.5	0	45.5	38.0	19.0	18	−1.0
(1)	twice mean =103	144.5	144.5	99.0	49.5	53	+3.5

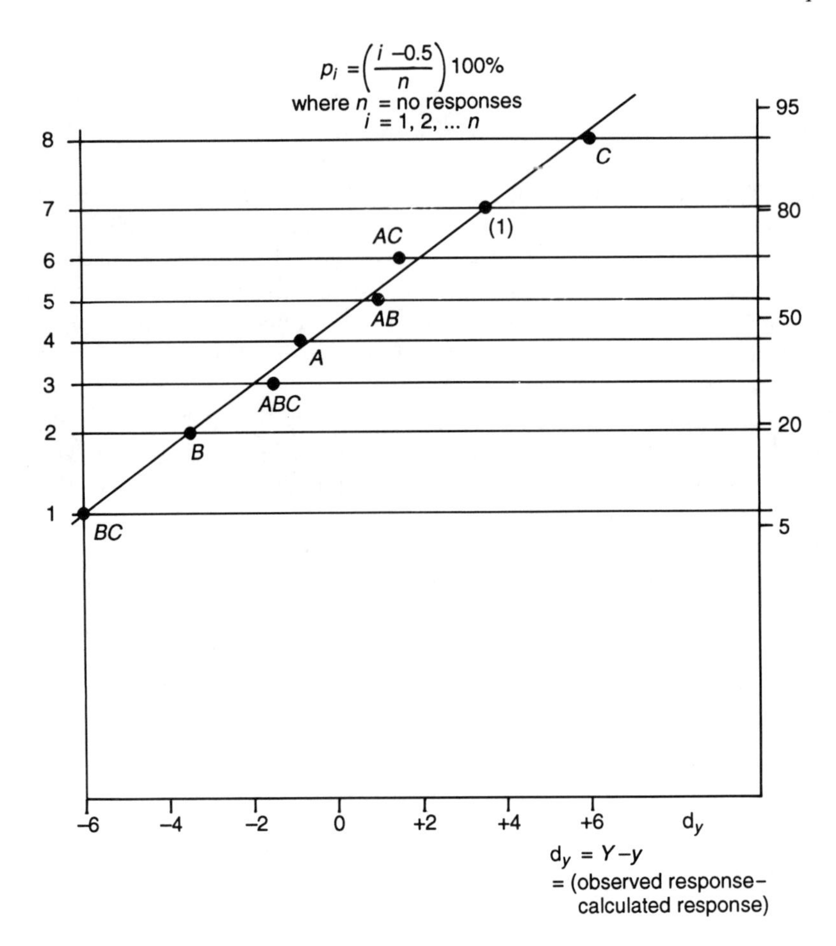

Figure 5.3 *Normal probability plot of residuals* d_y

Answer

The differences are now ABC -7, BC -0.5, AC -4, C $+11.5$, AB $+6.5$, B -9, A $+4.5$, (1) -2: a much poorer fit to a straight line. The model including AC is more satisfactory.

Discussion

It is not unusual for the two-factor interaction of significant variables itself to be significant. A good move now would be to carry out a 2^2 factorial in A and C, giving two further estimates each of eff_A, eff_C, and eff_{AC}.

Significant effects can therefore be identified from unreplicated factorials. Replication of the design is to be preferred, if circumstances permit, to give a directly observed estimate of the error variance. But what is to be done if, say, seven variables are to be investigated? This would lead to $2^7 = 128$ treatment combinations without duplication!

No-one need contemplate such a programme of work. A full 2^7 experiment would provide 64 estimates of the seven main effects – far more than one imagines are needed – and of 21 two-factor interactions (some of which may well be significant) and 35 three-factor interactions (where significance perhaps will occur); and 35 four-factor, 21 five-factor, seven six-factor interactions and one seven-factor interaction (all of which will be insignificant or anyway incomprehensible).

Correspondingly, many of the unit experiments would be a waste of effort. The only use of high-order interactions would be to obtain a pooled error estimate; but replication gives a much better estimate!

We shall see in Chapter 6 that it is possible to do only 32 unit experiments, *including* duplication, and still obtain estimates of the seven main effects not aliased with any first-order (two-factor) interactions, and of those 21 first-order interactions aliased in threes.

Now, 32 unit experiments, including duplication, clearly means a design of 16 treatment combinations – a 2^4. So what is proposed is only a one-eighth fraction of the full design. How can doing only a one-eighth replicate possibly give much information? To get the answer, it is necessary to understand and use the concepts of confounding and aliases. We shall see how to use confounding to divide a design into fractions to get information we want, at the expense of information we can probably or very probably do without. (Later we shall see how we can also get that other information if we need it.) We shall also see how the confounding we choose determines what is aliased with what, so that we can try to ensure that important effects are not aliased with one another.

5.8 SIGNIFICANCE OF ONE OR SEVERAL MEAN SQUARES IN FACTORIAL EXPERIMENTS

If in a single test an experimental F-value, F_e, is greater than (say) the 5% chance value, we infer that the factor or interaction concerned probably has a real effect. The chance that this inference is wrong is less than 5%.

However, in analysing a factorial experiment, numerous effects may be assessed, and each assessment will carry some uncertainty – some chance of an erroneous conclusion. For instance, if all fifteen mean squares of a 2^4 factorial are assessed, and even if no factor or interaction has a real effect, there is a chance of about one in two that at least one F_e will reach the 5% level.

Therefore, it is probably best to place little reliance on the 5% level of significance in such an analysis if only (say) one of six or more F-values just reaches the 5% value. The experiment should be made larger so as to make it more sensitive. If any mean square reaches the 1% level, this may be taken as sufficient evidence of the existence of a real effect.

This caution is particularly important if the conclusion that an effect is real will entail an expensive investment (in, say, plant design or modification).

REFERENCES

1. W.E. Duckworth, 'Statistical Techniques in Technological Research: an Aid to Research Productivity', Methuen, London, 1968, p. 60.
2. L. Davies, *Talanta*, 1975, **22**, 371–377.
3. M.J. Moroney, 'Facts from Figures', 2nd revised Edn., Penguin, Harmondsworth, 1970.
4. D.R. Cox, 'Planning of Experiments', Wiley, New York, 1958, Chapter 5.
5. Reference 1, pp. 135–138.
6. G.E.P. Box, W.G. Hunter, and J.S. Hunter, 'Statistics for Experimentalists', Wiley, New York, 1979, pp. 333–334, 344.
7. C. Daniel, 'Applications of Statistics to Industrial Experimentation', Wiley, New York, 1976, p. 74.
8. Reference 7, pp. 71–82.
9. 'The Design and Analysis of Industrial Experiments', ed. O.L. Davies, 4th Edn., Longman, London, 1978, pp. 95–96.
10. G.B. Weatherill, 'Intermediate Statistical Methods', Chapman and Hall, London, 1981, Chapter 8.
11. Reference 6, numerous references.
12. G.E.P. Box and N.R. Draper, 'Empirical Model-building and Response Surfaces', Wiley, New York, 1987, numerous references.

CHAPTER 6

Fractional Factorial Designs at Two Levels

We are now going to be even more economical in our designs than we were with factorial experiments: we shall divide a complete factorial design into fractions (blocks) with a view to doing only one or at most some of them. Before doing that, it is helpful to know all the treatment combinations of a full factorial without any feat of memory.

6.1 THE TOTAL TREATMENT COMBINATIONS IN A FACTORIAL

Every factorial design is said to have a set of $n+1$ multipliers. For example, in a 2^4 design with factors A, B, C and D, these are (1), a, b, c, and d.

To derive the list of TCs, start by writing (1). Now multiply it by a, yielding a. Having done everything you can with (1) and a, multiply through by b to give b and ab. Working similarly with c and d gives the full set of TCs:

(1), *a*, *b*, *ab*, *c*, *ac*, *bc*, *abc*, *d*, *ad*, *bd*, *abd*, *cd*, *acd*, *bcd*, *abcd*.

You will recognize the first eight symbols as those for a 2^3 design, and you can go on multiplying from the beginning by variables $E,F,G\ldots$. Hence you can derive any full set of TCs and in the order demanded by the Yates table.

Exercise 6.1

How many TCs will end with the symbol f, and which are the 8th, 14th, 21st, and 32nd in Yates table order?

Answer

32; *abcf*, *acdf*, *cef*, *abcdef*.

6.2 DIVIDING A FACTORIAL EXPERIMENT INTO BLOCKS

Suppose we have a 2^3 experiment which we divide into two blocks of four TCs. By doing so, we have introduced another source of possible variation – there may be some systematic difference between the blocks. But we have no spare degree of freedom to accommodate this, so the between-blocks difference must become confounded with one of the 2^3-1 effects of the 2^3 design. That is to say, the subdivision causes us to lose all knowledge of the magnitude of one of the effects. (If we divided into four blocks, we would lose knowledge of three effects, and so on.)

Clearly, we wish this to be the least important effect. In a 2^3 design, this is probably the *ABC* interaction, so this is the effect we choose to confound. (A word of caution here: this has led to the general impression that any set of confounded effects should include the highest-order interaction. But this is not sound advice when dividing into more than two blocks – see, *e.g.*, the example in Section 6.5.) Before we see what the effect of confounding *ABC* in a 2^3 design is, let us see what might go wrong if we choose our two blocks unwisely.

6.3 CONFOUNDING OF EFFECTS IN FACTORIAL EXPERIMENTS DIVIDED INTO BLOCKS

Consider a 2^3 experiment which is divided between two batches of raw material as follows:

Batch I	*Batch II*
a	(1)
ab	*b*
ac	*c*
abc	*bc*

and suppose batch I is superior to batch II, so that every experiment with it gives an extra yield X (which is an unknown magnitude).*

Without the extra yield, the results of the experiments would be

$$(1),\ a,\ b,\ ab,\ c,\ ac,\ bc,\ abc;$$

and

$$eff_A = \tfrac{1}{4}[a-(1)+ab-b+ac-c+abc-bc] = x$$

With the extra yield,

$$\begin{aligned} eff_A &= \tfrac{1}{4}[a+X-(1)+ab+X-b+ac+X-c+abc+X-bc] \\ &= x+X \end{aligned}$$

The effect of A, as observed, =(the true effect of A plus the effect of the difference between batches). Since we do not know the effect of that difference, we cannot know the true effect of A. We say that eff_A is *confounded* with the difference between batches.

We have no idea of the importance (if any) of factor A, because we have not randomly distributed the two batches across the two blocks. A similar confounding would occur if we allocated the first block (as given above) to operator X, and the other to operator Y: we would have confounded eff_A with the between-operators difference. Whenever an experiment is divided into p blocks, $p-1$ effects are

* If we *know* the batches are different, and we wish to investigate factors other than the difference between batches, we are *wise* to use one batch only per block (see Section 2.8 above).

confounded with between-blocks variance. To choose effects to confound (to lose), remember that significance is likely to get less as the interaction order increases. Hence the choice of ABC when dividing a 2^3 in two.

However, there are consequences of our choices. *When confounding a 2^n experiment into 2^p blocks, we choose p interactions to be confounded. Another (2^p-p-1) interactions – or factors if we choose badly – will also be confounded.* Hence, when dividing a 2^3 experiment into 2^1 blocks, we choose one interaction to confound; then, since $2^1-1-1=0$, there is no other confounding. If we choose ABC, *and do the whole experiment*, we lose only eff_{ABC}. If we do only one block, then aliasing occurs, and the last statement is no longer true (see Section 6.6). As another example, if we divide a 2^6 experiment into 2^3 blocks, we choose three interactions, and another $(2^3-3-1)=4$ effects will be confounded. In general, the total number of confounded factors or interactions is 2^p-1, the number of degrees of freedom between blocks.

6.4 RULE FOR SUBDIVISION OF A FACTORIAL EXPERIMENT INTO TWO BLOCKS

This rule depends upon the form of the effect equations. For instance, for a 2^3 experiment, the effect equation for eff_{ABC} is

$$eff_{ABC}=\tfrac{1}{4}[a+b+c+abc-ab-ac-bc-(1)]$$

The rule for subdivision of the 2^3 experiment into two blocks so as to confound ABC is: put all the positive terms of the ABC effect equation into one block, and all negative terms into the other. Hence:

Batch I	*Batch II*
(1)	a
ab	b
ac	c
bc	abc

Since all batch I experiments give an extra yield X, we now have

$$\begin{aligned} eff_{ABC}(\text{observed}) &= \tfrac{1}{4}\{a+b+c+acb-(ab+X)-(ac+X)-(bc+X)-[(1)+X]\} \\ &= eff_{ABC}(\text{true})-X \end{aligned}$$

Since X is unknown, the true eff_{ABC} cannot be measured; it is confounded with the between-batches (between-blocks) difference. But eff_A is now free of the effect of the batches:

$$eff_A(\text{observed})=\tfrac{1}{4}\{a+(ab+X)+(ac+X)+abc-b-c-(bc+X)-[(1)+X]\}$$

The Xs cancel, so $eff_A(\text{observed})=eff_A(\text{true})$.

Indeed, all effects except eff_{ABC} are clear of the difference between batches – check for yourself!

We can reach further conclusions if we assume that the true eff_{ABC} is zero (as it is often believed to be). If this is the case, the observed eff_{ABC} is a measure of the difference between batches plus any other difference(s) between blocks. This is why it can be used as the error estimate in the absence of replication or previous knowledge. If the observed eff_{ABC} is not significant compared with the experimental error obtained from replication, these differences are not significant compared with the experimental error.

It is not necessary to remember or look up the effect equations when subdividing by this method. Any effect equation can be factorized, having n factors for a 2^n design. For instance, in a 2^3 design

$$\begin{aligned} eff_A &= \tfrac{1}{4}[a+ab+ac+abc-b-c-bc-(1)] \\ &= \tfrac{1}{4}(a-1)(b+1)(c+1) \end{aligned}$$

In general, if a factor F is involved in an effect, the effect equation includes a term $(f-1)$. If F is not involved, the term is $(f+1)$. Hence, for ABC, we write

$$eff_{ABC} = \tfrac{1}{4}(a-1)(b-1)(c-1)$$

Expanding this gives the full equation which decided the subdivision into blocks.

Exercise 6.2

Derive the effect equation for eff_B in a 2^4 design.

Answer

$$\begin{aligned} eff_B &= \tfrac{1}{4}(a+1)(b-1)(c+1)(d+1) \\ &= b-(1)+ab-a+bc-c+bd-d+abc-ac+abd \\ &\quad -ad+bcd-cd+abcd-acd \end{aligned}$$

6.5 COMPLETE SETS OF CONFOUNDED EFFECTS

When we choose p effects to confound, and another (2^p-p-1) effects are also confounded, these additional effects are the 'generalized interactions' of the chosen ones. For example,

the generalized interaction of A and B is AB
the generalized interaction of AB and CD is $ABCD$
the generalized interaction of ABC and ACD is BD

This last example illustrates the general rule for *deriving generalized interactions* in *two-level* experiments: multiply factors/interactions and count any squared term as unity. This is known as multiplying to modulo 2.

Exercise 6.3

What are the generalized interactions of (i) *abcf.acdf.cef*, (ii) *cdg.degk.cek*?

Answer

(i) *bcdef*, (ii) (1).

So, if we have a 2^6 experiment with factors A, B, C, D, E, F, and we divide it into

2^3 blocks, we can choose *three* interactions to confound (say, $ABCDEF$, $ABCDE$, $ABCDF$). The chosen interactions are called the *defining contrasts* of the design. The other $(2^p - p - 1) = 2^3 - 3 - 1 = 4$ effects/interactions which are then confounded are:

$$ABCDEF.ABCDE = F$$
$$ABCDEF.ABCDF = E$$
$$ABCDE.ABCDF = EF$$
$$ABCDEF.ABCDE.ABCDF = ABCD$$

Note that this is a very poor design if E and F are important factors, since no information can be obtained about them. (And if they are not anticipated as possibly important, why include them in the design in the first place?) This is a clear example of the inadvisability of arbitrarily using highest and next-highest interactions as defining contrasts in multi-block designs – main effects become confounded. We have to choose interactions as defining contrasts, so that no main effect is confounded and the confounded interactions are probably of little importance or at least are known to us (see Section 6.9).

The reason why particular effects cannot be estimated from a factorial divided into blocks becomes clear when what is called the *principal block* of the design is worked out. If the principal block does not have half of its TCs at the high level of any variable X, and consequently the other half at low X, neither it nor any other block of the design can estimate eff_X (see Sections 6.6 and 6.7).

6.6 FRACTIONAL FACTORIAL EXPERIMENTS

These experiments are the most economical method of assessing the importance of a number of system variables and such of their interactions as we choose to investigate. As Duckworth[1] said in introducing fractional factorials in his book, 'Now you can really begin to save money and time'. Technological investigations commonly involve up to seven, ten, or even more possible factors. Fractional factorials at two levels are an excellent way to secure some major economizing while studying them.

This is achieved by dividing the full factorial into blocks, with the intention of completing only the experiments of one (or at most a few) of the blocks. While any block may be chosen at random, there are good reasons for doing the principal block first. The *principal block* is that block which includes TC(1).

Consider the 2^3 experiment divided into two blocks so that ABC is confounded (Section 6.4 above):

Block I (principal block)	*Block II*
(1)	*a*
ab	*b*
ac	*c*
bc	*abc*

Now the Yates table for the *full* 2^3 experiment shows that

$$eff_A = \frac{1}{4}[a + ab + ac + abc - b - c - bc - (1)$$

and

$$eff_{BC} = \frac{1}{4}[a + bc + abc + (1) - b - c - ab - ac]$$

If only Block I is done, we have only four results, and must write

$$eff_A = \frac{1}{2}[ab - (1) + ac - bc]$$
$$eff_{BC} = \frac{1}{2}[-ab + (1) - ac + bc]$$

So, under this circumstance, $eff_A = -eff_{BC}$, and eff_A and eff_{BC} are *aliases* of one another. In fact, the eff_A we measure is actually $eff_A - eff_{BC}$. However, interaction effects are usually smaller than main effects, so if only the principal block is done, the contribution of $-eff_{BC}$ is ignored. The better thing to do is to go on and do Block II; the two blocks together give separate estimates of eff_A and eff_{BC} (see Chapter 7). It is not advisable to do only a half-replicate of a 2^3 design, and this example is worked through here only as a simple illustration.

The complete set of aliases for a 2^3 in A, B, C divided by confounding ABC are given by the generalized interactions rule:

$A.ABC = A^2BC = BC$: A and BC are aliased
$B.ABC = AC$: B and AC are aliased
$C.ABC = AB$: C and AB are aliased

You may be surprised that I have just written that A and BC are aliased, which might make you think $eff_A = +eff_{BC}$, whereas a few lines earlier I derived the correct relationship *for the principal block*, $eff_A = -eff_{BC}$. Negative signs arise for the principal block whenever an effect is multiplied by an interaction which has an *odd* number of letters. Strictly speaking, an odd number defining contrast in a principal block has a negative sign (here, $-ABC$). Hence $A.-ABC = -BC$ and $eff_A = -eff_{BC}$. Multiplying by an *even* numbered interaction in the principal block leads to positive signs. Hence, *e.g.*, $CD.BCDF$ leads to $eff_{CD} = +eff_{BF}$, whereas $CD.BCDEF$ gives $eff_{CD} = -eff_{BEF}$. The sign rule for other blocks is given in Section 7.1.

Probably surprisingly at first sight, it is better to include more factors in a fractional design, for then it is easy to avoid confounding a main effect with a two-factor interaction, or one two-factor interaction with another. The *resolution*[2] of a two-level fractional factorial design is the length of the shortest confounded interaction. In general, a design of resolution r is one in which no p-factor effect is confounded with any other effect containing less than $r-p$ factors. The resolution is indicated by a roman letter subscript.

For example, we would designate a half-fraction of a 2^5 factorial with $ABCDE$ as its defining contrast as a 2_{V}^{5-1} design. The main factors would be confounded only with four-factor interactions when only this one block is done, and two-factor with three-factor interactions. A resolution IV design confounds main factors with three-factor interactions, and two-factor interactions with one another. If the shortest confounded interaction has only three letters, main effects are confounded with two-factor interactions. It is obviously better to have longer confounded interactions, which means more factors.

6.7 DERIVATION OF PRINCIPAL BLOCK AND SUBSEQUENT BLOCKS OF A FRACTIONAL FACTORIAL DESIGN

The standard method to derive the *principal* block is:

> Put into the principal block all TCs which have zero, 2, 4... letters in common with each of the defining contrasts.

A simple example is the 2^3 design with *ABC* confounded. Those effects which have zero or two letters in common with *ABC* are (I), *AB*, *AC*, *BC*. The principal block is (1), *ab*, *ac*, *bc*.

This method is cumbersome in larger designs. An alternative, systematic method of deriving the principal block from the design multipliers is explained in Section 6.10.

The *remaining* blocks of a design are derived by multiplying the contents of the principal block successively by elements not in that block nor in any block already formed by the multiplicative procedure. Therefore, the second block of the 2^{3-1} design is obtained by Block I $\times$ $a \rightarrow a$, *b*, *c*, *abc*.

Exercise 6.4

Divide a 2^4 factorial in *W*, *X*, *Y*, *Z* into two blocks by confounding *WXYZ*.

Answer

Block I: (1), *wx*, *wy*, *xy*, *wz*, *xz*, *yz*, *wxyz*
Block II: *w*, *x*, *y*, *wxy*, *z*, *wxz*, *wyz*, *xyz*

As another example, take the 2^6 experiment in 2^3 blocks, choosing *ABCDEF*, *ABCDE*, and *ABCDF* as defining contrasts (section 6.5 above): the TCs with 0, 2, or 4 letters in common with each of these interactions form the principal block. The other seven blocks arise as shown in Table 6.1.

Table 6.1 *Blocks from defining contrasts ABCDEF, ABCDE, ABCDF (a bad design)*

Block I	*II* =*I.a*	*III* =*I.e*	*IV* =*I.f*	*V* =*I.ae*	*VI* =*I.af*	*VII* =*I.ef*	*VIII* =*I.aef*
(1)	*a*	*e*	*f*	*ae*	*af*	*ef*	*aef*
ab	*b*	*abe*	*abf*	*be*	*bf*	*abef*	*bef*
ac	*c*	*ace*	*acf*	*ce*	*cf*	*acef*	*cef*
bc	*abc*	*bce*	*bcf*	*abce*	*abcf*	*bcef*	*abcef*
ad	*d*	*ade*	*adf*	*de*	*df*	*adef*	*def*
bd	*abd*	*bde*	*bdf*	*abde*	*abdf*	*bdef*	*abdef*
cd	*acd*	*cde*	*cdf*	*acde*	*acdf*	*cdef*	*acdef*
abcd	*bcd*	*abcde*	*abcdf*	*bcde*	*bcdf*	*abcdef*	*bcdef*

This full design makes clear the point mentioned in Section 6.5: no block has a mixture of levels of either E or F. Hence no estimate can be made of eff_E, eff_F.

A valid design to estimate each main factor will always have, in each block, half the TCs at high level (and the other at low, of course), for any main factor which appears in it.

6.8 ALIASING IN FRACTIONAL FACTORIALS

Whenever a factorial experiment is divided into 2^p blocks *and carried out completely*, $2^p - 1$ effects are confounded with the between-blocks difference. In the example just considered, the confounded effects are $ABCDEF$, $ABCDE$, $ABCDF$, E, F, EF, and $ABCD$.

When only one block is completed, additional information is lost. What actually happens is that *the effects corresponding to the TCs of that block become aliased by their generalized interactions with the confounded effects.*

This was introduced in Section 6.6. When we divide a 2^3 into two by confounding ABC we have the alias pairs A, BC; B, AC; C, AB. For the first three lines of the $2^6/2^3$ design in the last section, we have

Block I × confounded effects = aliases of Block I

Block I				Aliases			
(I)	$ABCDEF$	$ABCDE$	$ABCDF$	E	F	EF	$ABCD$
AB	$CDEF$	CDE	CDF	ABE	ABF	$ABEF$	CD
AC	$BDEF$	BDE	BDF	ACE	ACF	$ACEF$	BD

[If you want to know what (I) is, it is a non-existent or zero effect, corresponding to the mean of all the experimental results – look at the Yates table.]

You will notice from these few lines that this is a poor design, not only because E and F are lost, but because there is aliasing between two-factor interactions (from the examples given, $eff_{AB} = eff_{CD}$ and $eff_{AC} = eff_{BD}$).

6.9 SYSTEMATIC CHOICE OF DEFINING CONTRASTS

Sections 6.7 and 6.8 have shown that the allocation of TCs to blocks *and* the aliasing of effects depend on the choice of defining contrasts. When we carry out a block of a fractional factorial, the TCs we perform, and the aliases involved, are determined by this choice. Now, we must design our division into blocks so that effects we expect to be significant are aliased only by effects which we expect to be insignificant. *The selection of defining contrasts is* therefore *of prime importance.*

There are two ways of going about this. The first is to look into a table of published designs, which will identify certain sets of defining contrasts and outline the consequences – whether main effects and two-factor interactions are clear of one another, or whether there is confusion between them, and so on. The pioneering textbook 'Design and Analysis of Industrial Experiments' devoted its Chapter 10 to a discussion and table of nine fractional designs to assess between three and eight factors by between eight and sixteen observations. In each case, the

TCs of the principal block are given. Bisgaard[3] has produced a listing of 48 two-level full and fractional factorials, to deal with three to 15 factors in eight – or sixteen – run designs.

However, Greenfield[4] has stated that textbooks 'give no guidance as to the selection of defining contrasts with regard to the experimental objectives other than by trial and error'. He therefore proposed a systematic approach, which resulted in an improved method by Franklin and Bailey.[5] It is this latter procedure which is now discussed.

Example 6.1

Consider a 2^5 experiment in A, B, C, D, E. The smallest possible fractional design is required which will estimate each main effect and also the interaction effects AB and BE. All other interactions are considered negligible. Greenfield called the set of effects to be estimated the *requirements set* A, B, C, D, E, AB, BE. What we need to know is, how can we find the defining contrasts which will do this? The design must also estimate the mean; with the seven previous requirements, this requires a minimum of eight observations. Hence, of the $2^5(=32)$ possible TCs, we must choose at least $8(=2^3)$ suitable to the requirements, to form a quarter-replicate 2^5 design. We are setting out to design a 2^5 experiment in four (*i.e.* 2^2) blocks – a 2^{5-2} experiment. (In general, we can divide a 2^n experiment into 2^m blocks of 2^{n-m} TCs each. A general rule-of-thumb is appropriate here: it is unwise to design fractional factorials of less than eight TCs in multi-factor designs: aliasing becomes rampant.)

A 2^3 design will by now be familiar as a full factorial for three factors (say, A, B, C). In working out the principal block to analyse for the requirements set given above, these three are called the *basic factors*, while D and E are the *added factors*. (In general, we have $n-m$ basic factors and m added factors.) We begin our identification of appropriate defining contrasts by constructing a table with the eight effects of the full 2^3 placed vertically and the added factors placed horizontally (Table 6.2a). (Note that the vertical column follows the order for the Yates analysis.)

Table 6.2a *Structure of a Franklin–Bailey table*

		Added factors	
		D	*E*
	I		
	A		
	B		
Basic effects	*AB*		
	C		
	AC		
	BC		
	ABC		

Table 6.2b *Selection of defining contrasts by the Franklin–Bailey table for Example 6.1*

		Added factors		
		D	*E*	
	I	~~*D*~~	~~*E*~~	
	A	~~*AD*~~	~~*AE*~~	
	B	~~*BD*~~	~~*BE*~~	
Basic effects	*AB*	~~*ABD*~~	~~*ABE*~~	Ineligible effects are deleted
	C	~~*CD*~~	~~*CE*~~	
	AC	*ACD*	*ACE*	
	BC	*BCD*	~~*BCE*~~	
	ABC	*ABCD*	*ABCE*	

We now have to decide which effects we can choose as defining contrasts. For example, we clearly cannot use any of our required effects as a defining contrast; if we do, it will become totally confused with the between-blocks difference. We have to distinguish between *eligible effects* which may be chosen for defining contrasts, and ineligible effects which may not.

For a given requirements set, all the effects in that set are ineligible for selection as defining contrasts, and so are all generalized interactions between pairs in that set, and the mean effect *I*.

So, for the requirement set *A*, *B*, *C*, *D*, *E*, *AB*, *BE* the ineligible set is

I, *A*, *B*, *AB*, *C*, *AC*, *BC*, *ABC*, *D*, *AD*, *BD*, *ABD*, *CD*, *E*, *AE*, *BE*, *ABE*, *CE*, *BCE*, *DE*, *BDE*

Since we wish to divide the full factorial into four blocks, we must find two defining contrasts and their generalized interaction (which are not among the ineligible set). We do this by multiplying basic effects and added factors, using the usual rule that squared symbols = unity. In the present example, this results in Table 6.2b.

We now select the first eligible effect in the first added-factors column, *ACD*, and search the second column for any eligible effect whose generalized interaction with *ACD* is also eligible. Neither *ACD*.*ACE* = *DE* nor *ACD*.*ABCE* = *BDE* is eligible. We therefore move to the next eligible effect in the first column and search the second column. Here we have *BCD*.*ACE* = *ABDE* and *BCD*.*ABCE* = *ADE*. Both of these are eligible. Multiplying the final effect in the first column by *ACE* or *ABCE* does not produce an eligible generalized interaction (do you agree?) So we have two sets of three eligible effects *ACE*, *BCD*, *ABDE* and *ADE*, *BCD*, *ABCE*. Either may be used to divide the 2^5 design into 2^2 blocks to meet the requirements. Either of these will generate a principal block in which none of the required effects *A*, *B*, *C*, *D*, *E*, *AB*, *BE* is aliased – *i.e.* confused – with any other.

To estimate the magnitudes of these effects, we

(i) work out the principal block;

(ii) construct the aliasing matrix to see that no required effect is aliased with any other such effect. Since we are looking for seven effects from an eight-row Yates table, this means that there must be one required effect on each effect row. Two on a row means there is an aliasing between required effects;

(iii) carry out the principal block and analyse the results by Yates' table.

Choosing defining contrasts and deriving the principal block and alias matrix correctly is essential if we are to analyse unambiguously for the effects we want. Therefore not one but two exercises will follow for each of these activities.

Exercise 6.5

Eight main factors *A*–*H* are to be estimated from 16 TCs without any aliasing between them. Use the Franklin–Bailey method to show that *CDG*, *ABDF*, *ACDH*, and *BCDE* form an eligible set of defining contrasts.

Answer

Estimating eight factors from 16 TCs means choosing a 2^4 fraction from a 2^8 factorial. In this exercise, therefore, there will be four basic factors and four added

Table 6.3 *Partial Franklin–Bailey table for {A, B, C, D, E, F, G, H} requirements set in a 16 TC experiment*

	Added factors			
Basic effects	*E*	*F*	*G*	*H*
⋮				
ABD		*ABDF*		
CD			*CDG*	
ACD				*ACDH*
BCD	*BCDE*			
ABCD				

ones. The ineligible set is the eight factors and all the two-factor interactions which come from them, and *I*. The Franklin–Bailey table is, in part, as in Table 6.3.

Discussion

CDG, *ABDF*, *ACDH*, and *BCDE* is one eligible set of defining contrasts. They meet the necessary criteria:

(i) each added factor appears once in a defining contrast;
(ii) each chosen contrast comes from a different row of the table.

To summarize, each contrast must come from a different row and a different column of the Franklin–Bailey table.

Exercise 6.6

Use the Franklin–Bailey method to derive defining contrasts for a 2^4 fractional factorial which will estimate each of the seven interactions *AB*, *AC*, *AD*, *BC*, *BD*, *BE*, *BF* without aliasing between them, or with the main factors *A–F*. [Hint: since there are six factors, this problem is equivalent to deriving a 2^4 fraction from a full 2^6 factorial – *i.e.* one quarter of the full design. Dividing a design into four (2^2) blocks requires that you find two defining contrasts (and their generalized interaction). The need for two defining contrasts and the related difference of two in power numbers, 2^4 and 2^6, both mean that there will be two added factors. Set out the Franklin–Bailey table with basic effects *I–ABCD* and added factors *E*, *F*.]

Answer

Requirements set	{*AB AC AD BC BD BE BF*}
Ineligible set	{*I AB AC AD BC BE BF AE AF CD ABCD ABCE ABCF ABDE ABDF CE CF DE DF EF*}

There are two options for defining contrasts: *ACDE*, *BCDF* and *BCDE*, *ACDF*. The generalized interaction for each is *ABEF*.

Table 6.4 *Franklin–Bailey table for requirements set {AB AC AD BC BD BE BF}*

	Added factors	
Basic effects	*E*	*F*
I	*E*	*F*
A	—	—
B	—	—
AB	*ABE*	*ABF*
C	—	—
AC	*ACE*	*ACF*
BC	*BCE*	*BCF*
ABC	—	—
D	—	—
AD	*ADE*	*ADF*
BD	*BDE*	*BDF*
ABD	—	—
CD	*CDE*	*CDF*
ACD	*ACDE*	*ACDF*
BCD	*BCDE*	*BCDF*
ABCD	*ABCDE*	*ABCDF*

Discussion

The requirements set gives a Franklin–Bailey table as shown in Table 6.4.
However, one would never nominate a main factor as a defining contrast. Secondly, all the second-order interactions except *CDE*, *CDF* would alias a required interaction with a factor. Thirdly, taking *CDE*, *CDF* would alias *E* with *F*. This is not forbidden by this particular question, but would be a thing to avoid. Taking a fourth-order and third-order defining contrasts (bottom row of Table 6.4 with the first or second up) would give *AEF* or *BEF* as generalized interaction. *AEF* is permissible in this question, but is normally something to avoid, since it would alias a main factor with a first-order interaction. Similarly, using the bottom row with the fourth row up is permissible here, yielding *ABEF*, but the fourth-row choice would alias a factor with a two-factor interaction.

If we keep to four-factor confounded interactions, main factors will be aliased only with three-factor interactions, but there will be aliasing between two-factor interactions. There are two options in this question, *ACDE*, *BCDF* and *BCDE*, *ACDF*. The design will then be designated 2^{6-2}_{IV}.

6.10 IDENTIFICATION OF PRINCIPAL BLOCK FROM CHOSEN DEFINING CONTRASTS

The principal block can be separated from the total set of TCs by the standard method (TCs with 0, 2, 4 . . . letters in common with each of the defining contrasts). This method is tedious and error-prone, however, in larger designs. Fortunately, it is not necessary to use it. What we do instead is modify the multipliers which generate the basic effects of the Franklin–Bailey table, by multiplying them, appropriately, by the added factors. To find out what to multiply with, we look at the defining contrasts: see Example 6.2.

Example 6.2

Suppose we choose *ACE* and *BCD* as our defining contrasts from Example 6.1. In that case, the basic effects formed a 2^3 factorial with multipliers (1), *a*, *b*, *c*; the added factors were *D*, *E*. In these defining contrasts, *D* is associated with *B* and *C*, and *E* with *A* and *C*. Rearranging, *A* is associated with *E*, *B* with *D*, and *C* with *D* and *E*.

As a consequence, we multiply the multipliers of the basic effects, except (1), as follows:

(i) multiply *a* by *e*
(ii) multiply *b* by *d*
(iii) multiply *c* by *de*

In Example 6.1, therefore, the multipliers are (1), *ae*, *bd*, *cde*.

The TCs for the principal block are now obtained by multiplying the multipliers in the normal way. For example, instead of $a.b=ab$, we have $ae.bd=abde$; and $a.c.=ac$ becomes $ae.cde=acde^2=acd$. The principal block is derived as (1), *ae*, *bd*, *abde*, *cde*, *acd*, *bce*, *abc*. To check whether these are correct, use the rule that each should have zero or two letters in common with *ACE*, *BCD*: they have. To check that they are in Yates analysis order for a 2^3 design, bracket round the basic symbols: they are.

Exercise 6.7

The defining contrasts from Exercise 6.1 not used in Example 6.2 were *BCD*, *ABCE*. Show that choosing these leads to the principal block (1), *ae*, *bde*, *abd*, *cde*, *acd*, *bc*, *abce*.

Answer

In *BCD*, *ABCE*, *D* is associated with *BC* and *E* with *ABC*. Rearranging, *A* is associated with *E*, *B* with *D* and *E*, and *C* with *D* and *E*. The multipliers therefore are (1), *ae*, *bde*, *cde*. These give the principal block shown.

Exercise 6.8

A 2^7 factorial in *A–G* was divided into 2^3 blocks by choosing *ABCE*, *ABDF*, *ACDG* as defining contrasts. Show that the principal block is (1), *aefg*, *bef*, *abg*, *ceg*, *acf*, *bcfg*, *abce*, *dfg*, *ade*, *bdeg*, *abdf*, *cdef*, *acdg*, *bcd*, *abcdefg*.

Answer

Using *ABCE*, *ABDF*, *ACDG* means that, of the added factors, *E* is associated with *ABC*, *F* with *ABD* and *G* with *ACD*. Rearranging, *A* is associated with *EFG*, *B* with *EF*, *C* with *EG*, and *D* with *FG*. The multipliers are (1), *aefg*, *bef*, *ceg*, *dfg*, and they give the principal block shown.

6.11 YATES ANALYSIS AND ALIASING MATRICES IN FRACTIONAL FACTORIALS

The responses from a fractional factorial are subjected to Yates analysis exactly as in Tables 1.2 and 5.1. As illustration, consider the principal block derived from Example 6.1 in Example 6.2. Representing the responses by the TCs, we derive the estimates of eff_A, eff_B, ... eff_{ABC} as shown in Table 6.5.

But we are not seeking the effects of A, B, C and their interactions. We wish to estimate the effects of A, B, C, D, E, AB, and BE. One of these seven effects must be on each line of the Yates table [except the (1) line] if eight TCs are to give us the

Table 6.5 *Yates table for the principal block of a quarter-replicate 2^5 design with defining contrasts ACE, BCD*

	Columns of analysis		
Response	1	2	3
(1)	(1) + *ae*	(1) + *ae* + *bd* + *abde*	(1) + *ae* + *bd* + *abde* + *bce* + *abc* *cde* + *acd* = Total
ae	*bd* + *abde*	*bce* + *abc* + *cde* + *acd*	*abde* − *bd* + *ae* − (1) + *abc* − *bce* + *acd* − *ade* = $4eff_A$
bd	*cde* + *acd*	*abde* − *bd* + *ae* − (1)	*bce* + *abc* − *cde* − *acd* + *bd* + *abde* − (1) − *ae* = $4eff_B$
abde	*bce* + *abc*	*abc* − *bce* + *acd* − *cde*	*abc* − *bce* − *acd* + *cde* + *abde* − *bd* − *ae* + (1) = $4eff_{AB}$
cde	*ae* − (1)	*bd* + *abde* − (1) − *ae*	*bce* + *abc* + *cde* + *acd* − (1) − *ae* − *bd* − *abde* = $4eff_C$
acd	*abce* − *bd*	*bce* + *abc* − *cde* − *acd*	*abc* − *bce* + *acd* − *cde* − *abde* + *bd* − *ae* + (1) = $4eff_{AC}$
bce	*acd* − *cde*	*abde* − *bd* − *ae* + (1)	*bce* + *abc* − *cde* − *acd* − *bd* − *abde* + (1) + *ae* = $4eff_{BC}$
abc	*abc* − *bce*	*abc* − *bce* − *acd* + *cde*	*abc* − *bce* − *acd* + *cde* − *abde* + *bd* + *ae* − (1) = $4eff_{ABC}$

Table 6.6 *Alias matrix for a quarter-replicate 2^5 design principal block with confounded interactions ACE, BCD, ABDE*

		Confounded interactions	
Basic effects	*ACE*	*BCD*	*ABDE*
(*I*)	−*ACE*	−*BCD*	+*ABDE*
*A**	−*CE*	−*ABCD*	+*BDE*
*B**	−*ABCE*	−*CD*	+*ADE*
*AB**	−*BCE*	−*ACD*	+*DE*
*C**	−*AE*	−*BD*	+*ABCDE*
AC	−*E**	−*ABD*	+*BCDE*
BC	−*ABE*	−*D**	+*ACDE*
ABC	−*BE**	−*AD*	+*CDE*

* = wanted effect.

results we want. The *aliasing matrix* tells us where to place our wanted effects in the table. The matrix is derived by multiplying the effects for a full factorial in the basic factors – the basic effects for our fractional factorial – successively by the confounded interactions (not just the defining contrasts). This is done in Table 6.6.

The aliases are in the body of the table. The positive and negative signs arise from the rule for the principal block governed by the number of letters in the confounded interaction. We see that no required effect is aliased by another. We can estimate A, B, C, D, E, AB, BE from eight treatment combinations only. We have maximum efficiency in the sense that we have estimated seven effects from only eight experiments. However, as a 2^{5-2} design with three-letter shortest confounded interactions, this is only a resolution III design with main factors aliased with two-factor interactions.

Looking at the aliasing matrix (Table 6.6) and the right-hand column of the Yates table (Table 6.5), we re-write that column to read, from top to bottom,

$$4eff_A,\ 4eff_B,\ 4eff_{AB},\ 4eff_C,\ -4eff_E,\ -4eff_D,\ -4eff_{BE}$$

We can then use the Yates table to calculate the effects we want.

Exercise 6.9

A quarter-replicate 2^5 in A, B, C, D, G used the defining contrasts ACD, BCG. This design was used to study the effect of five factors on a dimerization reaction, where the response was the product ratio (=dimer/trimer + tetramer). The results were:

(1)	*ad*	*bg*	*abdg*	*cdg*	*acg*	*bcd*	*abc*
4.26	0.87	5.14	1.06	1.04	3.13	1.32	2.56

(i) Work out the alias matrix
(ii) Show that the effect estimates were

eff_A	eff_B	eff_C	eff_D	eff_G
−1.04	+0.19	−0.83	−2.70	+0.35

Answer

(i)

	ACD	*BCG*	*ABDG*
(*I*)	−*ACD*	−*BCG*	+*ABDG*
*A**	−*CD*	−*ABCG*	+*BDG*
*B**	−*ABCD*	−*CG*	+*ADG*
AB	−*BCD*	−*ACG*	+*DG*
*C**	−*AD*	−*BG*	+*ABCDG*
AC	−*D**	−*ABG*	+*BCDG*
BC	−*ABD*	−*G**	+*ACDG*
ABC	−*BD*	−*AG*	+*CDG*

(ii)

Factor	TC	Response	Columns of analysis 1	2	3	Effect $=\underline{3}/4$
	(1)	4.26	5.13	11.33	19.36	
A	*ad*	0.87	6.20	8.03	−4.16	−1.04
B	*bg*	5.14	4.17	−7.47	0.76	+0.19
	abdg	1.06	3.86	3.31	−1.56	−0.39
C	*cdg*	1.04	−3.39	1.07	−3.30	−0.83
$-D$	*acg*	3.13	−4.08	−0.31	10.78	+2.70
$-G$	*bcd*	1.32	2.09	−0.69	−1.38	−0.35
	abc	2.56	1.22	−0.87	−0.18	−0.05

[Notes. You may care to use the unused lines of the Yates table to give an estimate of error variance, which here is $[2(-0.39)^2+2(-0.05)^2/2]$, and do an ANOVA with $\phi_1=1$, $\phi_2=2$, to show that D is approaching highly significant and A and C are possibly significant on a one-tailed test. This is sometimes referred to in the literature as the 'method of dummy variables.' Or you may care to look back to your Daniel's plot (Exercise 4.1) to show that D, A, and C obviously look significant, and then show that the residuals on this basis are a reasonable fit to a straight line. The lower sensitivity of the ANOVA here is because the error term includes a relatively large effect, −0.39, and of course there are only two degrees of freedom to play with.]

Exercise 6.10

In 'The Design and Analysis of Industrial Experiments', O.L. Davies and his collaborators suggested a quarter-fraction of a 2^6 factorial in A–F by using $ABCE$, $BCDF$ as defining contrasts. (i) Show that the principal block is (1), *ae*, *bef*, *abf*, *cef*, *acf*, *bc*, *abce*, *df*, *adef*, *bde*, *abd*, *cde*, *acd*, *bcdf*, *abcdef*. (ii) Use the alias matrix to show that each factor is aliased with two three-factor and one five-factor interactions, and that the two-factor interactions are aliased in pairs except in one case ($AE=BC=DF$). (iii) Also use the matrix to show the Yates table lines on which eff_E and eff_F appear.

Answer (in part)

(ii) $A\times ABCE,\ BCDF,\ ADEF\rightarrow BCE,\ ABCDF,\ DEF$
$BC\times ABCE,\ BCDF,\ ADEF\rightarrow AE,\ DF,\ ABCDEF$

(iii) $E\times ABCE\rightarrow ABC$; so $F=+ABC$, and appears on line 8.
$F\times BCDF\rightarrow BCD$; so $F=+BCD$, and appears on line 15.

6.12 UNDERSTANDING THE VALIDITY OF FRACTIONAL FACTORIALS

The question might be asked, 'How can $abde-bd+ae-(1)+abc-bce+acd-cde$ from Table 6.5 represent $4eff_A$, when previously it was $ab-b+a-(1)+abc-bc+ac-c$?'.

The answer comes in two parts:

(i) Remember that what we get are only estimates of eff_A; we do not expect absolute truth from either.

(ii) However, having said that, look more closely at the first expression. We can bracket the added factors and write it as $ab(de)-b(d)+a(e)-(1)+abc-bc+ac(d)-c(de)$. What remains unbracketed is the second expression for eff_A. Note that two comparisons for eff_A are at high D ($abde-bd$ and $acd-cde$) and two at low D [$ae-(1)$ and $abc-bce$]. Also, two comparisons are with E falling as A falls [$abde-bd$ and $ae-(1)$] while in the others, E rises as A falls ($abc-bce$ and $acd-cde$). In these ways, the effects of D and E on A are balanced out. If the original assumption of negligible AD, AE interactions is valid, there are no such effects anyway.

Of course, we accept the existence of aliases when we do fractional factorials, but we shall see in Section 7.1 how aliased effects can be separated.

6.13 FAILURE OF AN INITIAL FRACTIONAL DESIGN

It was shown in Section 6.11 that A, B, C, D, E, AB, BE can be estimated from only eight TCs. However, it can be shown that if the requirements set is modified to A, B, C, D, E, AB, **CE**, it is not possible to do only a 2^3 block without aliasing between required effects. The Franklin–Bailey table is as in Table 6.2b, but the previously eligible ACE, $ABCE$ in the E column are now ineligible, as they would alias both A and AB with CE. There is thus no valid entry in the E column. Correspondingly, there can be only one added factor, and the full design can only be divided into two blocks: *i.e.* $2^4=16$ TCs must be done. For a division into two, use the highest interaction $ABCDE$: a2_{V}^{5-1} design.

REFERENCES

1. W.E. Duckworth, 'Statistical Techniques in Technological Research', Methuen, London, 1968, p. 96.
2. G.E.P. Box, W.G. Hunter, and J.S. Hunter, 'Statistics for Experimentalists', Wiley, New York, 1979, p. 385.
3. S. Bisgaard, 'A Practical Aid for Experimenters', Starlight Press, 1010 Starlight Drive, Madison, WI 33705, USA.
4. A.A. Greenfield, *Appl. Stat.*, 1976, **25**, 64–67.
5. M.F. Franklin and R.A. Bailey, *Appl. Stat.*, 1977, **26**, 321–326.

CHAPTER 7

Fractional Factorial Designs in Sequence

There are considerable advantages in dividing a multifactor design into fractions (blocks) and carrying them out, selectively, one at a time. Aliased effects can be separated from one another, undesirable or insignificant variables can be quickly dropped, modified levels or new variables can be introduced, and transformations of input and response variables can be carried out if necessary.

Some writers, such as the Japanese engineer Genichi Taguchi, have recommended quite the reverse procedure: namely, beginning with a comprehensive experiment in which every factor is to be investigated. However, suppose that we do indeed carry out a full two-level factorial and find that one factor has a significant adverse effect on the desired result. One half of the TCs will have been at the undesirable level, and time, effort, money, and materials will have been wasted. Only the other half of the design is useful for assessing the other factors. If only a fraction of the total design is done initially, the loss is reduced in direct proportion. We shall therefore now look at some of the advantages of a planned sequence of fractions, and the fractions can if necessary build up to a full factorial design. Additional arguments, in favour of this sequential approach, are put forward in Chapter 12, where this philosophy is further compared with that of Professor Taguchi.

7.1 REMOVING ALIAS AMBIGUITIES

If we divide a factorial design into p blocks, each effect estimated by doing one block is aliased with $(p-1)$ other effects. It follows that if we divide into fewer blocks, each effect has fewer aliases. For instance, if we carry out a one-quarter design, each effect has three aliases. If we then do another of the blocks, the total is half the complete factorial, and each effect has only one alias. This reduction in the number of aliases is called the *separation of aliases*. The ambiguity posed by 'alias strings' must usually be resolved by such a sequence of fractions (or, occasionally and less convincingly, by an appeal to technical knowledge). *The block to separate any factor X from its two-factor aliases is the principal block multiplied by x.*

Combining the results from the principal block and that block, so as to secure alias separation, requires the correct algebraic signs for the aliases, which derive from getting the signs for the defining contrasts right for each block.

7.1.1 Rule for Transforming Defining Contrasts Algebraic Signs from Principal Block to Any Other Block

You will remember from Section 6.7 that we transform the principal block into another block by multiplying it by any effect not within it. The rule about the defining contrasts signs is: *change the sign of any defining contrast once for each letter in common between that contrast and the multiplying effect.*

This arises from the way effect equations are divided up when we divide a full factorial into blocks. For example, it was shown in Section 6.6 that the principal block of a 2^{3-1} factorial confounding ABC has $eff_A = -eff_{BC}$. The defining contrast in that block is therefore $A.-BC=-ABC$. The other half of the 2^3 is derived by multiplying the principal block by a (or b or c). In it, $eff_A = 1/2(a-b+abc-c) = eff_{BC}$; so the defining contrast is $A.BC=+ABC$. If, instead, we multiply by any two-factor interaction, we come back to the principal block. The sign changes when there is an odd number of letters in common, but is unchanged when there is an even number in common.

Example 7.1

ACE and BCD were chosen as defining contrasts in Example 6.2 to carry out a 2^{5-2} principal block. In that block, ACE and BCD, being odd-numbered, are actually $-ACE$, $-BCD$. If any or all of factors A, B, and C were found significant from that block, they can be separated from their first-order aliases by carrying out the appropriate blocks:

Block II = principal block $\times a$, for which the defining contrasts are $+ACE$, $-BCD$

Block III = principal block $\times b$, where the defining contrasts are $-ACE$, $+BCD$

Block IV = principal block $\times c$, with defining contrasts $+ACE$, $+BCD$.

(Note that when all the blocks are listed, all plus–minus combinations of the signs are included. If not, something is wrong.) The generalized interactions follow: principal block $+ABDE$, block II $-ABDE$, block III $-ABDE$, block IV $+ABDE$.

Exercise 7.1

For a 2^{4-1} design confounding $ABCD$, work out (i) the principal block, and (ii) block II = the principal block $\times d$. Expand the effect equations $eff_D = (a+1)(b+1)(c+1)(d-1)$ and $eff_{ABC} = (a-1)(b-1)(c-1)(d+1)$. Therefrom, work out the effect equations for D and ABC in (iii) the principal block, and (iv) block II. From these, show that (v) $eff_D = +eff_{ABC}$ in the principal block, and (vi) $eff_D = -eff_{ABC}$ in block II. Hence, demonstrate that the rule about signs holds for even-numbered as well as odd-numbered defining contrasts.

Answer

(i) (1) *ad bd ab cd ac bc abcd*

(ii) *d a b abd c acd bcd abc*

(iii) $\mathit{eff}_D = \frac{1}{4}[ad-(1)+bd-ab+cd-ac+abcd-bc]$
$\mathit{eff}_{ABC} = \frac{1}{4}[ad-(1)+bd-ab+cd-ac+abcd-bc]$
(iv) $\mathit{eff}_D = \frac{1}{4}(d-a+abd-b+acd-c+bcd-abc)$
$\mathit{eff}_{ABC} = \frac{1}{4}(-d+a-abd+b-acd+c-bcd+abc)$
(v) From (iii), $\mathit{eff}_D = +\mathit{eff}_{ABC}$: defining contrast $+ABCD$
(vi) From (iv), $\mathit{eff}_D = -\mathit{eff}_{ABC}$: defining contrast $-ABCD$.

7.1.2 Separation of Aliases

If we write $\mathit{eff}_{A(O)}$ for the observed eff_A, and $\mathit{eff}_{A(S)}$ for eff_A separated from its aliases (our best estimate of the true eff_A) it follows in Example 7.1 that, *e.g.*

$\mathit{eff}_{A(O)} = x_1 = \mathit{eff}_{A(S)} - \mathit{eff}_{CE} - \mathit{eff}_{ABCD} + \mathit{eff}_{BDE}$ from the principal block
$\mathit{eff}_{A(O)} = x_2 = \mathit{eff}_{A(S)} + \mathit{eff}_{CE} - \mathit{eff}_{ABCD} - \mathit{eff}_{BDE}$ from block II.

If we ignore any interactions above two-factor, we may write, respectively,

$$x_1 = \mathit{eff}_A - \mathit{eff}_{CE}$$
$$x_2 = \mathit{eff}_A + \mathit{eff}_{CE}$$

and therefore

$$\mathit{eff}_A = \tfrac{1}{2}(x_1 + x_2)$$
$$\mathit{eff}_{CE} = \tfrac{1}{2}(x_2 - x_1)$$

Similarly, from the principal block and block III, $\mathit{eff}_B = \frac{1}{2}(x_3 + x_4)$, $\mathit{eff}_{CD} = \frac{1}{2}(x_4 - x_3)$. From the principal block and block IV,

$$\mathit{eff}_{C(O)} = x_5 = \mathit{eff}_{C(S)} - \mathit{eff}_{AE} - \mathit{eff}_{BD}$$
$$\mathit{eff}_{C(O)} = x_6 = \mathit{eff}_{C(S)} + \mathit{eff}_{AE} + \mathit{eff}_{BD}$$

and so

$$\mathit{eff}_C = \tfrac{1}{2}(x_5 + x_6),\ \mathit{eff}_{AE} + \mathit{eff}_{BD} = \tfrac{1}{2}(x_6 - x_5)$$

The main effects are thereby estimated free from their first-order interactions.

We are not limited to separating just the basic effects, because the design in Example 7.1 is a valid one for five variables from four blocks. In the defining contrasts, A is associated with E, and B with D. If you work out additional blocks of the design by multiplying the principal block by a, b, c, d, and e, you will find that the block obtained by multiplying by a is the same as that when multiplying by e. Consequently, that one block can separate both A and E from their two-factor interactions; and similarly the block obtained by multiplying by b or d separates both B and D.

Exercise 7.2

A 2^{6-3} experiment to investigate the effect of six variables A–F on the fluidity of cement, had the treatment combinations (1), *adf*, *bef*, *abde*, *cdef*, *ace*, *bcd*, *abcf* in Yates order.

(i) Work backwards from (1), $a(df)$, $b(ef) \ldots c(def) \ldots$ to show that D is associated with AC, E with BC and F with ABC; that the defining contrasts are therefore ACD, BCE, $ABCF$; and consequently the generalized interactions are $ABDE$, BDF, AEF, $CDEF$.

(ii) Use the three-factor confounded interactions to show the two-factor aliases of the six main factors in this principal block.

(iii) Factors A and C were found to have significant effects. Ignoring any aliases above two-factor, give the equations to relate the observed eff_A and eff_C to the best estimates from that block.

(iv) Derive the blocks II and III to separate A and C from their first-order aliases.

(v) If $eff_{A(O)} = +12.0$ (principal block), $+10.2$ (block II)
$eff_{C(O)} = +22.5$ (principal block), $+25.2$ (block III)
calculate $eff_{A(S)}$, $eff_{C(S)}$, and the sums of the two-factor interactions.

Answer

(ii) $A = -CD, \ -EF$; $B = -CE, \ -DF$; $C = -AD, \ -BE$, $D = -AC \ -BF$; $E = -AF, \ -BC$; $F = -AE, \ -BD$

(iii) $eff_{A(O)} = eff_{A(S)} - eff_{CD} - eff_{EF}$
$eff_{C(O)} = eff_{C(S)} - eff_{AD} - eff_{BE}$

(iv) Block II: *a, df, abef, bde, acdef, ce, abcd, bcf*
Block III: *c, acdf, bcef, abcde, def, ae, bd, abf*

(v) $eff_{A(S)} = 11.1$, $eff_{C(S)} = 23.85$, $eff_{CD} + eff_{EF} = -0.9$, $eff_{AD} + eff_{BE} = 1.35$.

7.1.3 Fold-over Designs

The method just described, of one successive block per significant variable, is not a particularly economical procedure (although we shall see it has another very useful application in Section 7.2). If all that is required is to separate all main factors from two-factor interactions, all we need to do is a particular second fraction. This is derived from the principal block by multiplying it by *all* the factors, and the two fractions together constitute a 'fold-over' design.[1]

The second block derived from the principal block in Exercise 7.2 would therefore be *abcdef, bce, acd, cf, ab, bdf, aef, de*. Every time we multiply a confounded interaction with a factor, we change its sign. Three-factor interactions will change sign three times: nett, a change of sign. Every such interaction will therefore change sign from the principal block to the fold-over block, and therefore every two-factor alias sign will change. The sum of the $eff_{X(O)}$ values from the two blocks will therefore be equal to twice $eff_{X(S)}$, where X is any factor. We must be very careful about alias signs when we calculate these effects and their associated interaction effects: see Example 7.2.

Example 7.2

A 2_{III}^{7-4} principal block with defining contrasts ABD, ACE, BCF, $ABCG$ and the corresponding fold-over block gave results as follows:

(1)	*adeg*	*bdfg*	*abef*	*cefg*	*acdf*	*bcde*	*abcg*
61	38	61	12	57	26	49	57
def	*afg*	*beg*	*abd*	*cdg*	*ace*	*bcf*	*abcdefg*
84	88	87	69	107	72	92	69

A Yates analysis, of the principal block only, gives the following estimates:

Effect	*A*	*B*	*AB*	*C*	*AC*	*BC*	*ABC*
Estimate	−23.8	−0.8	+3.3	+4.3	+12.3	+12.3	+16.3

The defining contrasts and their generalized interactions include the following three-factor effects: *ABD, ACE, AFG, BCF, BEG, CDG, DEF*. The two-factor aliases of the main factors are therefore as in Table 7.1.

The principal block estimates can therefore be re-written as in Table 7.2.

Table 7.1 *Two-factor aliases from defining contrasts ABD, ACE, BCF, ABCG*

	Principal block	*Fold-over block*
A	−*BD*, −*CE*, −*FG*	+*BD*, +*CE*, +*FG*
B	−*AD*, −*CF*, −*EG*	+*AD*, +*CF*, +*EG*
C	−*AE*, −*BF*, −*DG*	+*AE*, +*BF*, +*DG*
D	−*AB*, −*CG*, −*EF*	+*AB*, +*CG*, +*EF*
E	−*AC*, −*BG*, −*DF*	+*AC*, +*BG*, +*DF*
F	−*AG*, −*BC*, −*DE*	+*AG*, +*BG*, +*DE*
G	−*AF*, −*BE*, −*CD*	+*AF*, +*BE*, +*CD*
	(and +*ABC*)	(and −*ABC*)

Table 7.2 *Effect estimates, Example 7.2, principal block*

Estimate (x_1)	*Effect*
−23.8	*A*−*BD*−*CE*−*FG*
−0.8	*B*−*AD*−*CF*−*DG*
+3.3	*AB*+*CG*+*EF*−*D*
+4.3	*C*−*AE*−*BF*−*DG*
+12.3	*AC*+*BG*+*DF*−*E*
+12.3	*BC*+*AG*+*DE*−*F*
+16.3	−*AF*−*BE*−*CD*+*G*

Note carefully the algebraic signs in the effect column. For example *G* (*plus G*) appears on the *ABC* line because the even-numbered *ABCG* is a defining contrast: *ABCG.G*=*ABC*. But the confounded interactions *AFG, BEG, CDG* correspond to −*G*: Hence *ABC*= +*G*, −*AF*, −*BE*, −*CD*. *D, E*, and *F* appear with negative signs on the *AB, AC, BC* lines because of the odd-numbered confounded interactions *ABD, ACE, BCF*. But, *e.g.*, *ABCG* and *ABEF* are also confounded interactions. As they are even-numbered, *AB*= +*CG*= +*EF* on the *AB* line. So we write *AB*+*CG*+*EF*−*D*; and similarly for *E* and *F*.

Discussion of results, principal block

It looks as if *A, E, F* and *G* may all be important; but each main effect is aliased with three two-factor interactions. Significant factors may produce significant interactions; for example, A and G may produce an interaction big enough to make a substantial contribution to the estimate of its alias F (see Table 7.1).

It is therefore desirable to separate the factors from the interactions, by doing the fold-over block. Yates analysis of that block gives the estimates shown in Table 7.3.

All the algebraic signs in Table 7.3 are positive. In 'folding over', the three-factor confounded interactions have changed to positive signs. The interaction $ABCG$ will have changed signs four times, and will therefore still be positive: hence the bottom line of Table 7.3 refers to $ABC = +G$ *etc.*

We now pull the two sets of estimates together, using the expressions $eff_X = \frac{1}{2}(x_I + x_{II})$ and $eff_{\text{interactions}} = \frac{1}{2}(x_{II} - x_I)$ *paying careful attention to algebraic signs.* For example, look at the D line in Tables 7.2 and 7.3. The estimates are $x_I = 3.3 = -D + AB + CG + EF$ and $x_{II} = -2.5 = +D + AB + CG + EF$. Changing signs in the first equation gives $-x_I = -3.3 = D - AB - CG - EF$. Then, our final estimates are $eff_D = \frac{1}{2}(-3.3 - 2.5) = -2.9$, and $eff_{AB} + eff_{CG} + eff_{EF} = \frac{1}{2}[x_{II} - (-x_I)] = +0.4$.

The full results are in Table 7.4.

Table 7.3 *Effect estimates, Example 7.2, fold-over block*

Estimate (x_{II})	*Effect*
−18	$+A+BD+CE+FG$
−8.5	$+B+AD+CF+EG$
−2.5	$+AB+CG+EF+D$
+3	$+C+AE+BF+DG$
−11	$+AC+BG+DF+E$
−0.5	$+BG+AG+DE+F$
+8.5	$+AF+BE+CD+G$

Table 7.4 *Final estimated effects from principal and fold-over blocks, Example 7.2*

$eff_A = \frac{1}{2}(-18 - 23.8) = -20.9$	$eff_{BD+CE+FG} = \frac{1}{2}(-18 + 23.8) = 2.9$
$eff_B = \frac{1}{2}(-8.5 - 0.8) = -4.7$	$eff_{AD+CF+EG} = \frac{1}{2}(-8.5 + 0.8) = -3.9$
$eff_C = \frac{1}{2}(+3 + 4.3) = +3.7$	$eff_{AE+BG+DG} = \frac{1}{2}(+3 - 4.3) = -0.7$
$eff_D = \frac{1}{2}(-2.5 - 3.3) = -2.9$	$eff_{AB+CG+EF} = \frac{1}{2}(2.5 + 3.3) = +0.4$
$eff_E = \frac{1}{2}(-11 - 12.3) = -11.7$	$eff_{AC+BG+DF} = \frac{1}{2}(-11 + 12.3) = +0.7$
$eff_F = \frac{1}{2}(-0.5 - 12.3) = -6.4$	$eff_{BC+AG+DE} = \frac{1}{2}(-0.5 + 12.3) = +5.9$
$eff_G = \frac{1}{2}(+8.5 + 16.3) = +12.4$	$eff_{AF+BE+CD} = \frac{1}{2}(+8.5 - 16.3) = -3.9$

Discussion of final results

The main factors have been separated from their two-factor aliases, which remain aliased in threes. A remains the most important factor, with its aliases making only a small contribution to the responses. The situation with E and G is relatively similar. Compare these with F: compared with the first block result, about half the estimate has been parcelled out to the aliases by considering both blocks together. Perhaps this is a measurable AG interaction, as suggested earlier. A 2^2 factorial in A, G would now be a sensible thing to try. eff_B was nearly cancelled by the interaction effects, but is not a major effect; nor are C and D.

Exercise 7.3

A 2_{III}^{5-2} principal block with DFG, EFH as defining contrasts, and the corresponding fold-over block, were used to estimate the factors D, E, F, G, H.

(i) Derive the TCs for the two blocks

(ii) Derive the alias matrix to show that the factor effects appear on the following lines of the Yates table:

Factor	D	E	F	G	H
Line	2	3	5	6	7

(iii) If the effect estimates on lines 6 and 7 of the principal block are x_{16}, x_{17} respectively, and on the corresponding lines of the fold-over block are x_{26}, x_{27} respectively, show that the separated estimates for G, DF, H, and EF are
$eff_G = \frac{1}{2}(x_{26} - x_{16})$ $\quad eff_{DF} = \frac{1}{2}(x_{26} + x_{16})$
$eff_H = \frac{1}{2}(x_{27} - x_{17})$ $\quad eff_{EF} = \frac{1}{2}(x_{27} + x_{17})$

(iv) If the estimated effects were $x_{16} = +4.20$, $x_{26} = -11.60$, $x_{17} = +1.15$, $x_{27} = +8.95$, calculate eff_G, eff_{DF}, eff_H, eff_{EF}.

Answer

(i) (1) *dg eh degh fgh dfh efg def*
gh dh eg de f dfg efh defgh (both in Yates order)

(ii) Aliases:

Yates line	1	2	3	4	5	6	7	8
Basic effects	(1)	D*	E*	DE	F*	DF	EF	DEF
Basic effects × DFG	DFG	FG	$DEFG$	EFG	DG	G*	DEG	EG
Basic effects × EFH	EFH	$DEFH$	FH	DFH	EH	DEH	H*	DH

The required effects are on the lines as shown by the asterisks.

(iii) In the principal block $G.-DFG = -DF$ and $H.-EFH = -EF$; hence $eff_{DF} = x_{16} = eff_{DF} - eff_G$ and $eff_{EF} = x_{17} = eff_{EF} - eff_H$. In the fold-over block, $G.DFG = +DF$ and $H.EFH = +EF$; hence $eff_{DF} = x_{26} = eff_{DF} + eff_G$ and $eff_{EF} = x_{27} = eff_{EF} + eff_H$.
Hence, $eff_G = 0.5(x_{26} - x_{16})$, $eff_{DF} = 0.5(x_{26} + x_{16})$, $eff_H = \frac{1}{2}(x_{27} - x_{17})$, $eff_{EF} = \frac{1}{2}(x_{27} + x_{17})$.

(iv) $eff_G = -7.9$, $eff_{DF} = -3.7$, $eff_H = +3.9$, $eff_{EF} = +5.05$.

7.2 ESTIMATION OF THE TWO-FACTOR INTERACTIONS OF A SIGNIFICANT VARIABLE

When a factor has a large effect, it will very probably form appreciable interactions with other factors. (It can easily be shown[2] that, even with only two factors, the contours of the response surface would have to be *exactly* parallel to one factor, or *exactly* circular, for one factor's effect to be completely independent of the other factor. The chance of such a precisely zero interaction circumstance is very small, particularly as the number of factors increases.) It is therefore very helpful that the method of successive blocks not only separates a given variable from its two-factor aliases, but also allows the *estimation of the two-factor interactions of any chosen variable with all the others*: see Example 7.3.

Example 7.3

In Exercise 7.2, the principal block had $-ACD$, $-BCE$, $-BDF$, $-AEF$ as three-factor confounded interactions, together with $+ABCF$, $+ABDE$, $+CDEF$. The effect estimates may be set out as in Table 7.5 (ignoring aliases above first-order).

Since A was judged to be a significant factor, a second block (the principal block $\times\ a$) was carried out. For this second block, the confounded interactions are $+ACD$, $+AEF$, $-BCE$, $-BDF$, $-ABCF$, $-ABDE$, $+CDEF$. Again ignoring aliases above first-order, the effect estimates are as in Table 7.6.

Pairing corresponding results from the two blocks gives Table 7.7.

Table 7.5 *Estimates of effects from Exercise 7.2, principal block*

Yates order	*TC*	*Effect*	*Effect estimate*
(1)	(1)		
$a \to A$	*adf*	$A-CD-EF$	x_{11}
$b \to B$	*bef*	$B-CE-DF$	x_{12}
$ab \to AB$	*abde*	$AB+CF+DE$	x_{13}
$c \to C$	*cdef*	$C-AD-BE$	x_{14}
$ac \to -D$*	*ace*	$AC+BF-D$	x_{15}
$bc \to -E$*	*bcd*	$BC+AF-E$	x_{16}
$abc \to +F$*	*abcf*	$F-AE-BD$	x_{17}

*The identification of rows and algebraic signs for the added factors arises from the defining contrasts $-ACD \to D = -AC$; $-BCE \to E = -BC$; $+ABDF \to F = +ABC$.

Table 7.6 *Estimates of effects, principal block* $\times$ *a, Exercise 7.2*

Yates order	*TC*	*Effect*	*Effect estimate*
(1)	*a*		
$a \to A$	*df*	$A+CD+EF$	x_{21}
$b \to B$	*abef*	$B-CE-DF$	x_{22}
$ab \to AB$	*bde*	$AB-CF-DE$	x_{23}
$c \to C$	*acdef*	$C+AD-BE$	x_{24}
$ac \to +D$	*ce*	$AC-BF+D$	x_{25}
$bc \to -E$	*abcd*	$BC-AF+E$	x_{26}
$abc \to -F$	*bcf*	$F-BD+AE$	x_{27}

Table 7.7 *Estimates of the effects of factor A and its two-factor interactions, Exercise 7.2*

Effect	*Estimate*
A	$\frac{1}{2}(x_{11}+x_{21})$
AB	$\frac{1}{2}(x_{13}+x_{23})$
AC	$\frac{1}{2}(x_{15}+x_{25})$
AD	$\frac{1}{2}(x_{24}-x_{14})$
AE	$\frac{1}{2}(x_{27}-x_{17})$
AF	$\frac{1}{2}(x_{16}-x_{26})$
(also $CD+EF$	$\frac{1}{2}(x_{21}-x_{11})$)

Table 7.8 *Estimates of effects, principal block × c, Exercise 7.2*

Yates order	*TC*	*Effect*	*Effect estimate*
(1)	*c*		
a	*acdf*	$A+CD-EF$	x_{31}
b	*bcef*	$B+CE-DF$	x_{32}
ab	*abcde*	$AB-CF+DE$	x_{33}
c	*def*	$C+AD+BE$	x_{34}
ac	*ae*	$D+AC-BF$	x_{35}
bc	*bd*	$E-AF+BC$	x_{36}
abc	*abf*	$-F+AE+BD$	x_{37}

Table 7.9 *Estimates of the effects of factor C and its two-factor interactions, Exercise 7.2*

Effect	*Estimate*
C	$\frac{1}{2}(x_{14}+x_{34})$
AC	$\frac{1}{2}(x_{15}+x_{35})$
BC	$\frac{1}{2}(x_{16}+x_{36})$
CD	$\frac{1}{2}(x_{31}-x_{11})$
CE	$\frac{1}{2}(x_{32}-x_{12})$
CF	$\frac{1}{2}(x_{13}-x_{33})$
(also $AD+BE$	$\frac{1}{2}(x_{34}-x_{14})$)

Thus the two successive blocks not only separate variable A from its two-factor aliases, they also estimate the two-factor interactions of A with each of the other variables. The procedure could obviously be repeated for any other suspected significant variable.

Exercise 7.4

Combine the principal block in Exercise 7.2 with that block × c, to give the estimates of the two-factor interactions of C with each of the other factors.

Answer

The estimates from the second block are as in Table 7.8, since the confounded interactions are now $+ACD$, $-AEF$, $+BCE$, $-BDF$, $-ABCF$, $+ABDE$, $-CDEF$.

Pairing corresponding results gives Table 7.9.

7.3 GENERAL PRINCIPLES OF SEQUENTIAL FRACTIONS

Let us summarize the procedure to follow when a number of factors are believed to be important after initial screening. Proceed as follows:

(i) divide the design into blocks confounding higher-order interactions
(ii) carry out blocks one at a time (randomizing within each block)
(iii) analyse each block as soon as it is completed.

This is advisable even if a full factorial may ultimately be required for thorough study, because:

(i) the first block may give all the information required
(ii) one or more factors may give a large effect, causing all further work to be confined to the more-favourable levels of such factors
(iii) the experiment may be re-designed with fewer factors, perhaps at new levels, or, if desired, with other factors to replace ones dropped at the first trial.

(Note: this procedure assumes that when a factor shows a large effect, it will do so to some extent or other in all conditions of the other factors – *i.e.* interactions are not likely to appear, in later stages, which are so large as to obscure the effect noted earlier. When large interactions do appear, they indicate that the optimum may be near at hand, and that the process models obtainable from 2^n designs are inadequate – see Example 11.2 and Exercise 11.2.)

7.3.1 When to Stop Doing Successive Blocks

Continue the experiment until:

(i) the main effects are given with sufficient precision
(ii) important aliases between factors and two-factor interactions (or between pairs or sets of two-factor interactions) have been separated
(iii) no additional factors appear necessary to give adequate explanation of the results.

Increase of precision is obtained by replication of a given block, or by performing other block(s) of the same design. De-aliasing can be achieved by successive blocks, as just seen, or by carrying out small factorials in the factors of aliased interactions. Fitting data to models which satisfy criterion (iii) can take numerous forms.[3–5] The simple method of plotting estimates and estimate residuals on normal probability paper was described in Sections 4.1, 5.6, and 5.7, and the subject is returned to in Chapter 11.

REFERENCES

1. G.E.P. Box, W.G. Hunter, and J.S. Hunter, 'Statistics for Experimentalists', Wiley, New York, 1979, pp. 396–397, 426–429.

2. W.E. Duckworth, 'Statistical Techniques in Technological Research', Methuen, London, 1968, p. 132.
3. 'Statistical Methods in Research and Production', ed. O.L. Davies and P.L. Goldsmith, 4th revised Edn., Longman, London, 1978, pp. 264ff.
4. Reference 1, numerous references.
5. G.E.P. Box and N.R. Draper, 'Empirical Model-building and Response Surfaces', Wiley, New York, 1987, numerous references.

CHAPTER 8

Consistency of Processes and Products

In this book so far, the emphasis has been on whether one mean is significantly different from another, whether a mean effect is likely to be a significant one, and whether we can significantly improve a mean by some variation of a process. However, there is little point in improving a mean if the new process produces an unacceptably variable product. We must assure ourselves not only that the mean value of an effect is significant, but also that variability around that mean is not too large. This variability of an effect, a product, or a process (its reproducibility or *consistency*) is measured by the variance of replicate results about their mean. As we already know, variances can be compared by the F-ratio; hence we can use the F-test to compare consistencies.

Example 8.1

Consider the ten alloy samples of Table 2.4 in the test–retest case (Example 2.5). Is the heat-treated alloy significantly more variable than the untreated material?

Answer

See Table 8.1.

The F-test in Table 8.1 shows that the consistency of the alloy is not significantly altered by the heat treatment.

Exercise 8.1

Consider two sets of data which have standard deviations $s_1 = 0.0361$, $s_2 = 0.0173$, with $n_1 = 16$, $n_2 = 40$. Is there significant difference of consistency between the two sets?

Answer

$F_{15,39} = (0.0361)^2/(0.0173)^2 = 4.35$: $\alpha < 0.002$, two-tailed: highly significant difference. (These were the data of Exercise 2.11, where the significantly different variances precluded the use of the t-test for the means.)

Table 8.1 *Variability of tensile strength of alloy samples*

	Untreated		*Treated*	
Specimen	*Strength* x_1	$(x_1-x_1)^2$	*Strength* x_2	$(x_2-\bar{x}_2)^2$
1	20.7	0.0025	21.9	0.1849
2	18.4	5.5225	20.8	2.3409
3	19.8	0.9025	21.1	1.5129
4	18.8	3.8025	20.1	4.9729
5	19.9	0.7225	19.9	5.9049
6	23.4	7.0225	24.4	4.2849
7	23.7	8.7025	25.5	10.0489
8	20.8	0.0025	21.6	0.5329
9	20.0	0.5625	24.6	5.1529
10	22.0	1.5625	23.4	1.1449
	$\bar{x}_1=20.75$	$\Sigma(x_1-x_1)^2$ $=28.8050$	$\bar{x}_2=22.33$	$\Sigma(x_2-x_2)^2$ $=36.0810$
Estimate of population variance		28.8050/9 $=3.201$ $(\phi_1=9)$		36.0810/9 $=4.008$ $(\phi_2=9)$

Therefore $F_e=4.008/3.201=1.252$
From the F-table, $F_{9,9}=2.440$ for $\alpha=0.10$, single-sided.

8.1 CONSISTENCY IN FACTORIAL EXPERIMENTS

Up to now, we have summed the results for individual TCs in replicated experiments, in order to estimate the mean effects of factors and interactions on the responses. We now wish to estimate the effects of the factors and interactions on the consistency of the responses. Hence, we need an estimate of the variance of the responses at each TC.

Imagine that a factorial experiment has been carried out in duplicate. It can be shown that the variance of any duplicate pair of results x_1, x_2 about their mean:

$$\frac{\left(x_1-\frac{(x_1+x_2)}{2}\right)^2+\left(x_2-\frac{(x_1+x_2)}{2}\right)^2}{2-1}$$

reduces to $(x_1-x_2)^2/2$.[1] In a duplicate 2^3 experiment, we will have four estimates of the response variance for each effect at its high level, and four at its low level. The ratio of the means of these variances $=F_{4,4}$ (each estimate has one degree of freedom). If the F-value for an effect is significant, it means that factor or interaction significantly affects the consistency of the results.

In investigations which require high precision, such as in the validation of analytical methods, it may be necessary to repeat the experiment a number of times. If n sets of duplicate results are produced, the variance is $[\Sigma(x_1-x_2)^2]/2n$, and F will have $\phi_1=\phi_2=4n$.

The calculation can be done by Yates table, taking the responses as the square of the difference within any pair of duplicate results, $(x_1-x_2)^2$. There is no need to

Table 8.2 *Results of a 2^3 duplicate factorial*

TC	(1)	*x*	*y*	*xy*	*z*	*xz*	*yz*	*xyz*
Run 1	77.9	87.2	91.8	88.0	89.3	87.5	87.9	91.1
Run 2	79.5	74.8	99.0	92.6	85.3	94.5	90.1	91.5
Totals	157.4	162.0	190.8	180.6	174.6	182.0	178.0	182.6

divide by 2 (or $2n$) since this division cancels out in calculating F. The F-ratio for any effect E is determined by looking at the final analysis column. Add the Total figure to that for E, and divide the answer by the Total figure minus that for E (see Example 8.2 and footnote*).

If F calculated in this way is greater than unity, the variance is greater at the high level of any factor E. If it is fractional, the variance is greater at the low level of E. In the latter case, test the inverse of the experimental F to see if this low-level variance is significantly greater. Interpretation of an interaction effect is illustrated in Example 8.2.

Example 8.2

A 2^3 duplicate factorial in X, Y, Z gave the results shown in Table 8.2. The aim was to maximize the mean and minimize the variance.

The *mean effects* eff_E from the totals, and the corresponding mean squares ($MS_E = SS_E = 4eff_E^2$) are:

	X	Y	XY	Z	XZ	YZ	XYZ
Effect	+0.8	+7.0	−2.2	+3.3	+2.2	−6.0	+1.5
MS_E	2.6	196	19.4	40.6	19.4	144	9.0

* The justification for calculating F in this way is as follows. Table 1.2 shows that the figure in the last column of analysis opposite (say) *a* is equal to (all responses with A high − all responses with A low), while the figure at the top of that column, the Total, is (all responses with A high + all responses with A low). In Table 8.3, these figures are all 2(variance) values, so

$$\text{(all values with } A \text{ high} - \text{all values with } A \text{ low)} = 2(\text{variance due to } eff_A) = 2var_A$$

$$\text{(all values with } A \text{ high} + \text{all values with } A \text{ low)} = 2(\text{total variance}) = 2var_T$$

Hence,

$$F_A = \frac{2var_T + 2var_A}{2var_T - 2var_A}$$

$$= \frac{\text{values with } (A \text{ high}) + (A \text{ low}) + (A \text{ high}) - (A \text{ low})}{\text{values with } (A \text{ high}) + (A \text{ low}) - (A \text{ high}) + (A \text{ low})}$$

$$= \frac{\not{2} \text{ values with } A \text{ high}}{\not{2} \text{ values with } A \text{ low}}$$

From the sixteen individual results, coded by deducting 80.0 from each, total $SS = 583.5$ ($\phi = 15$); total treatment $SS = 431.0$ ($\phi = 7$); and so remainder $SS = 152.5$ ($\phi = 8$), giving residual variance $= 19.1$. The $F_{1,8}$ values are therefore $F_X < 1$; $F_Y = 10.3$; $F_{XY} \sim 1$; $F_Z = 2.1$; $F_{XZ} \sim 1$; $F_{YZ} = 7.6$; $F_{XYZ} < 1.0$. These correspond to Y and YZ having significant effects on the mean. Increasing Y increases the mean response. Also, because eff_{YZ} is negative, this beneficial Y effect will be greater at low Z. We therefore choose high Y, low Z.

We calculate the effects on the *consistency* of the results (the *variance effects*) as shown in Table 8.3.

We then calculate the F-values for the variance effects as follows:

	α single-sided
$F_X = \dfrac{299.6 + 149.1}{299.6 - 149.1} = 2.98$	NS (NS = not significant)
$F_Y = \dfrac{299.6 + (-143.6)}{299.6 - (-143.6)}$: inverted $= 2.84$	NS
$F_{XY} = \dfrac{299.6 + (-219.7)}{299.6 - (-219.7)}$: inverted $= 6.50$	<0.05 SIG
$F_Z = \dfrac{299.6 + (-159.6)}{299.6 - (-159.6)}$: inverted $= 3.28$	NS
$F_{XZ} = \dfrac{299.6 + (-92.5)}{299.6 - (-92.5)}$: inverted $= 1.89$	NS
$F_{YZ} = \dfrac{299.6 + 23.6}{299.6 - 23.6} = 1.17$	NS
$F_{XYZ} = \dfrac{299.6 + 144.3}{299.6 - 144.3} = 2.86$	NS

The α-values arise from tables for $F_{4,4}$:

F	4.11	6.39	9.60
(single-sided)	0.10	0.05	0.025

Note that Y and YZ, which significantly affected the mean, have no such effect on the variance. XY just about qualifies, on these results, as significantly affecting the consistency of the results ($\alpha < 0.05$) although its effect on the mean was insignificant. eff_{XY} is negative. Now, we wish to work at high Y to improve the mean. Doing so will make the variance effect of X more negative (the product consistency will be improved). Although eff_X in Table 8.3 was not significant, it was estimated as positive. eff_Y was also not significant, but was negative: the consistency was better at high Y. If these were real variance effects, and not just noise, it will be beneficial to work at low X and high Y. The total decision is low X, high Y, low Z. The marginal significance of XY would make further experimentation desirable, particularly if product variability would be a costly matter.

Table 8.3 *Calculation of variance effects in a duplicated 2^3 factorial*

TC	$\lvert x_1-x_2\rvert$	$\lvert x_1-x_2\rvert^2$ $\equiv$ *variance estimate*	1	2	3	
(1)	1.6	2.56	156.6	229.6	299.6	Total
x	12.4	154	73.0	70.0	+149.1	*X*
y	7.2	51.8	65.0	120.8	−143.6	*Y*
xy	4.6	21.2	5.0	28.3	−219.7	*XY*
z	4.0	16.0	151.4	−83.6	−159.6	*Z*
xz	7.0	49.0	−30.6	−60.0	−92.5	*XZ*
yz	2.2	4.84	33.0	−182	+23.6	*YZ*
xyz	0.4	0.16	−4.68	−37.7	+144.3	*XYZ*
		$\Sigma=299.6$				

(*Note*: the figures in the final analysis column are twice the correct estimate of variance, due to the omission of the divisor 2.)

We have seen in this Example how we can take advantage of knowledge about interactions. If we had obtained similar results, but by a method which did not estimate interactions, we would have identified only that *Y* had an important effect on the mean, and would have had no information as to how to reduce the variability of the product. We can take advantage in other ways, too: see Chapters 9 and 11.

Exercise 8.2

A duplicate 2^2 experiment in *A*, *B* gave differences between duplicates as follows:

	(1)	*a*	*b*	*ab*
x_1-x_2	6	20	5	3

Show that *A* is not significant in affecting consistency, but that *B* and perhaps *AB* are possibly significant. Suggest what you would do next.

Answer

TC	$(x_1-x_2)^2$	1	2
(1)	36	436	470
a	400	34	348
b	25	364	−402
ab	9	−16	−380

Hence, $F_A=818/122=6.7$; F_B (inverted) $=872/68=12.8$; F_{AB} (inverted) $=850/90=9.4$. From tables $F_{2,2}=9.0$, 19.0 for α (single-sided) $=0.10$, 0.05 respectively. Therefore, $F_A\ \alpha>0.10$, $F_B\ 0.10>\alpha>0.05$, $F_{AB}\ \alpha\sim0.10$. Replication of the design is desirable to increase precision. Note from these figures – and there are numerous others in this book – that a big response difference such as $a-(1)$ does *not* necessarily mean that eff_A will be important. What you have to remember are the effect equations. Here, $\mathit{eff}_B = (25-36) + (9-400)/2 = -201$, whereas $\mathit{eff}_A = (400-36) + (9-25)/2 = +174$.

REFERENCE

1. W.E. Duckworth, 'Statistical Techniques in Technological Research', Methuen, London, 1968, pp. 70, 185.

CHAPTER 9

The Optimization of Processes and Products

It has been shown in earlier chapters that 2^n factorial and fractional factorial experiments are economical and efficient ways of identifying and estimating the effects of changing the levels of system variables on the outcomes of chemical processes. It was also pointed out that when they identify significant interactions between variables, they suggest the presence, inside or outside the bounds of the experiment, of maxima or minima of the results to be expected from further experimentation. If the interaction is noticeably greater than the main factor effects, the maximum or minimum is relatively near at hand.

If such maxima or minima are suspected inside the original design, they can be looked for at factor levels between those originally chosen. This was the situation illustrated by Figure 1.1 and is indicated by declining (alternatively, increasing) responses when levels outside the original factorial are tested. The other possibility is that a maximum or minimum (or conceivably more than one) should be looked for outside the original design. This is indicated by responses not rising (or falling) as rapidly outside the factorial as would have been expected from results within it – see Figure 9.1.

In chemistry, of course, we often seek to optimize our processes to achieve such maxima (of yield or conversion of raw material) or minima (of impurity, process cost, or product variability). We have seen in Chapters 5 to 8 how factorial and fractional factorial experiments have great power to assist us in this search.

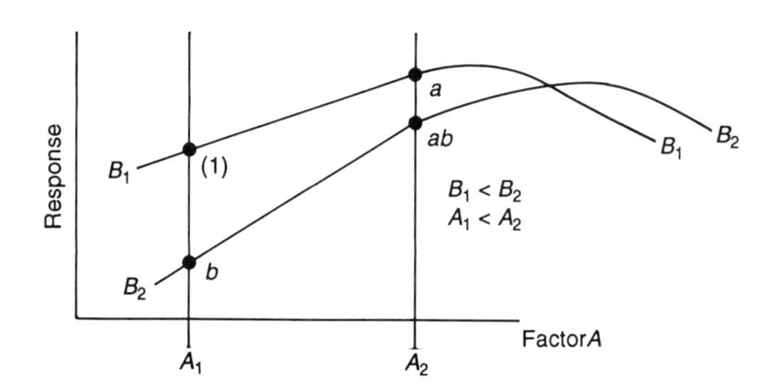

Figure 9.1 *AB interaction, indicating a maximum outside a 2^2 factorial design*

Alternatively, we can use two techniques called *method of steepest ascent* and *evolutionary operation*. The former method, due to Box and Wilson,[1] uses the magnitudes of the factor effects from a factorial or fractional factorial experiment to calculate new treatment combinations so as to search for the optimum. Evolutionary operation, also introduced by Box,[2,3] is not dependent on the prior availability of factorial results, although the original idea stemmed from factorial design.

Box intended EVOP to be a scheme for deliberately perturbing a working manufacturing process to obtain information which will lead to subsequent process or product improvement. The aim is to vary operating conditions sufficiently for the consequences to be assessed and yet not sufficiently far for those consequences to be appreciably harmful to output, quality, or safety. It is likely that changes of factor levels can only be small, and that it will be necessary to repeat a given set of experiments (a *phase*) a number of times before systematic changes of response become apparent. Box and his colleagues used 2^2 and 2^3 factorials, sometimes with a centroid, as the essential units of their EVOPS. An approach requiring fewer TCs, leading to the ultimate in experimental economy, is that of *simplex EVOP*. In this, after an initial set of TCs, unit experiments are designed and carried out (usually) one at a time. This is therefore the most dynamic approach, but one has to accept that it is less informative. The aim is simply to find the optimum without wondering why.

9.1 THE METHOD OF STEEPEST ASCENT

Example 9.1a

Suppose a 2^2 factorial in X, Y gave the following results: (1) 30, x 35, y 34, xy 39. These results correspond to $eff_X = 5$, $eff_Y = 4$. The method of steepest ascent now devises new TCs where the increase in the level of X is greater than that in the level of Y by the ratio eff_X/eff_Y; $= 5/4$ in this instance. Hence, if $(X_2 - X_1) = 10\,°C$ and $(Y_2 - Y_1) = 6$ atm, subsequent *steps* (the distance between any two neighbouring TCs in the ascent) would be 10 °C and $4/5(6) = 4.8$ atm, or suitable multiples or fractions thereof. How far away the new TCs are from the old depends on the experimenter, but the ratio of the increments is always the ratio of the effects. We can illustrate this Example by Figures 9.2 and 9.3. The steepest ascent originates at the centroid of the original design.

If the (unknown) response surface of yield is as in Figure 9.3, the new TCs would give yields (say) 47, 57, 52, and would encourage new experimentation round the highest yield just obtained. This will either include the optimum or indicate that better results lie beyond it.

You will have noted that the steepest ascent has been calculated from main effects only. In the simple example given, eff_{AB} is in fact zero, but in general interactions are ignored in the calculation, although they may well have triggered the steepest ascent approach in the first place. The calculation of the new TCs can incorporate allowance for significant interactions, but the calculation becomes

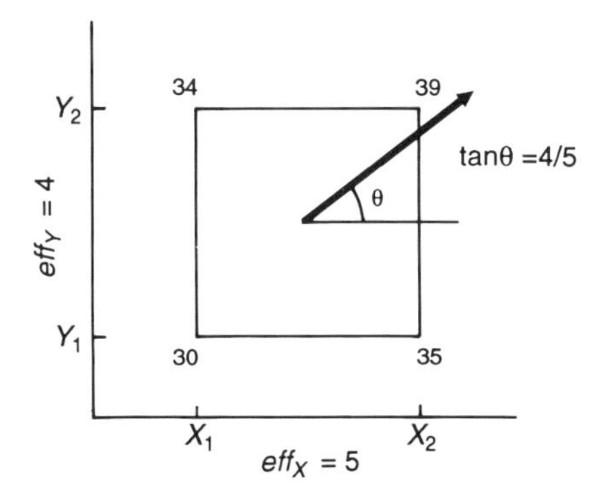

Figure 9.2 *Direction of steepest ascent from* $eff_X : eff_Y = 5:4$

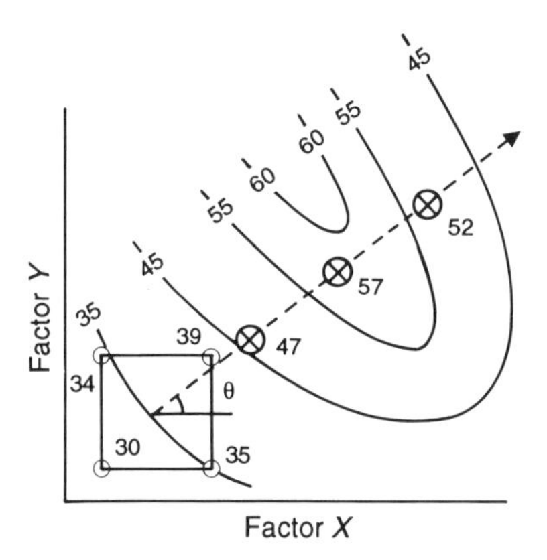

Figure 9.3 *Steepest ascent results*

complicated in multivariate systems (and makes unwarranted assumptions that the interactions near the optimum remain as they were within the original factorial). It is therefore usual to adopt the approach illustrated above – use main effects only as a sufficiently reliable guide to give a provisional optimum which can be tested by a further factorial, or by a *composite design* in which the new best TC becomes the centroid for a new factorial to which 'star points' are added (see Section 11.4).

Example 9.1b

A 2^3 factorial experiment in variables A (T: 80, 110 °C), B (time of addition of reagent: 1, 2 h), and C (acidity: 2.0, 5.0 M H^+) gave main effect estimates as follows: $eff_A - 11.5$, $eff_B = -14.5$, $eff_C = -1.0$, $eff_{AB} = -1.0$, $eff_{AC} = +0.5$, $eff_{BC} = -1.5$, $eff_{ABC} = 0$. Six steepest-ascent steps from the centroid were devised as follows. It was assumed that six steps in the most important factor, B, should equal the difference between its levels in the original experiment.

Note that *decreasing* B increases the yield. The difference in B levels was one hour; hence its step size will be 10 min, and the values will decrease. We must also *decrease* A to increase the yield. There will be six steps in A between 80 and 110 °C, but the steepest ascent steps will be only 11.5/14.5 of the B step size. Hence the A step size will be $11.5/14.5 \times (110-80)/6 = 4$ °C (very nearly). C also showed a (small) negative effect: reduce C also. The step size for C will be $1/14.5 \times (5-2)/6 = 0.03$ M (very nearly). The mid-point of the design was A 95 °C, B 1.5 h, C 3.5 M.

The steepest-ascent points are therefore

		New Tcs					
Variables	*Centroid*	1	2	3	4	5	6
A	95	91	87	83	79	75	71
B	1 h 30 min	1.20	1.10	1.00	0.50	0.40	0.30
C	3.5 M	3.47	3.44	3.41	3.38	3.35	3.32

For the important variables, A and B, only the fourth to sixth new TCs are outside the original design. If (as is often the case) the search is outside it, only these TCs would normally be done. The calculation has ignored the interactions. But eff_{AB}, eff_{AC}, and eff_{BC} were only comparable to the small eff_C, and eff_{ABC} was zero. Omitting the interaction effects should make little difference in such a case.

If interaction effects are more substantial, observed responses may soon begin to diverge noticeably from predicted values (see Section 9.1.1) as the ascent or descent proceeds. If, as our investigation develops, interactions cannot be ignored, then quadratic effects are not negligible, and we must estimate them (see Chapter 10 and particularly Chapter 11).

Exercise 9.1

A 2^2 factorial in D, E was carried out to develop a steepest ascent to improve the breaking strain of a fibre. The results were as shown:

D	90	100	90	100
E	40	40	50	50
Breaking strain	297	156	227	126

Devise eight steepest ascent steps from the centroid, assuming five steps for each variable between the original design limits, and assuming that the minimum permitted value of D is 80.

Answer

$$eff_D = -121, \; eff_E = -50 \; (\text{and } eff_{DE} = +20)$$

The step for D will be $0.2(100-90) = 2$ units; that for E will be $50/121 \times 0.2(50-40) = 0.83$ units.

The steps will be

		New TCs							
Centroid		1	2	3	4	5	6	7	8
D	95	93	91	89	87	85	83	81	79
E	45	44.17	43.34	42.51	41.68	40.85	40.02	39.19	38.36

TCs 3–8 are outside the original design for the most important factor, D. Because of its greater importance, TCs from 3 onwards would be done, even though only 7 and 8 are outside the limits for E. However, 8 would be below the permitted minimum for D, so 3–7 would be done.

9.1.1 Predicting Responses

In a case such as Example 9.1b, where the interaction affects are small or very small, it is possible to assume that the effects of the main factors continue linearly outside the factorial. We can then calculate future results, because each step should affect the response in a predictable way.

In Example 9.1b, factor B caused an increase of 14.5 units from high to low B, so the increase per step (of the six steps across the factorial) should be $14.5/6 = 2.42$ units. The steepest-ascent steps are the same size for B, so if we took eight steps from the centroid, B should increase the result by $8 \times 2.42 = 19.36$ units.

Factor A caused an increase of 11.5 units from high to low A; the increase per step (across the factorial) $= 11.5/6 = 1.92$ units. But the steepest-ascent steps of A would be only 11.5/14.5 the size; so we predict that eight A steepest-ascent steps will increase the response by $8(11.5/14.5)1.92 = 12.18$ units.

C had a very small effect (1 unit in 6 steps) and its steepest-ascent step will be only 1/14.5 times those in the original design. So that estimated increase due to C in eight steps will be $8(1/14.5)(1/6) = 0.10$.

The total increase would therefore be predicted as $19.36 + 12.18 + 0.1 = 31.64$ units above the result at the centroid. The centroid result is taken as the mean response from the factorial (which would be precisely true if all interactions were zero). When interactions are not negligible, the prediction is more uncertain; therefore uncertainty increases as maxima or minima are approached.

9.2 EVOLUTIONARY OPERATION

9.2.1 Box EVOPS

The evolution of a Box EVOP can be illustrated by starting with Figure 9.2. Since an EVOP is a development of an existing process, the four treatment combinations shown there would be arranged symmetrically around the existing conditions of that process – *i.e.* their centroid. The four experiments shown in Figure 9.2 would be called the first phase of the EVOP. Having done them one or more times, a new set of four would be carried out, placed symmetrically around the best result of phase 1 (Figure 9.4). This second phase usually includes the original process conditions, to check that it is not, after all, the best way of operating.

This procedure of building up phases around the successive best points is repeated until no further improvement is obtained and the results begin to deteriorate. If two results in a phase are equally good, the next stage has only two experiments, to produce a phase which is a 'mirror image' of the previous one (see the derivation of phase 3 from phase 2 in Figure 9.4). It is also permissible to do only three experiments, as in phase 5, although the cautious would also repeat

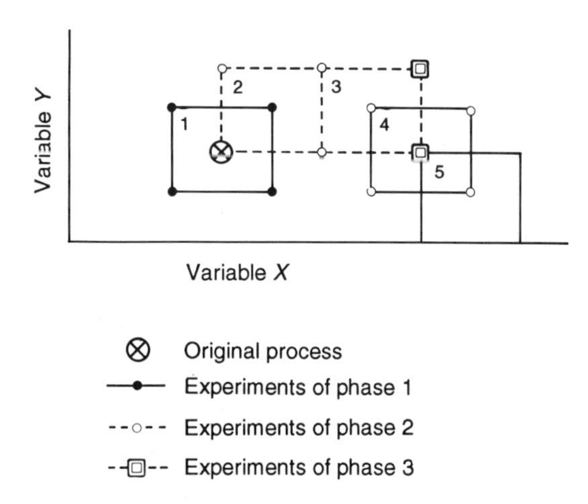

Figure 9.4 *A Box-type evolutionary operation for two variables X,Y*

experiments which are common between phases.

As each cycle is a complete factorial, a Yates analysis can be done at any stage to re-assess the ascent direction; or ANOVA can be done if there is replication.

A three-factor Box EVOP can be represented graphically by a succession of cubes which have edges, part-edges, part-faces (and conceivably full faces) in common. Development in more than two factors can be done by a numerical tabulation method similar to that illustrated in Section 9.2.4 below. But ensuring that the necessary changes are made from run to run becomes complex as the number of factors rises, and simplicity of operation may dictate that factors are studied in different combinations two at a time.

9.2.2 Simplex EVOPS

Box EVOPS use up to four to eight experiments per phase to assess the effect of altering the levels of two to three variables. But the most economical designs always estimate n effects from $n+1$ TCs. Therefore, most economically, one can seek to examine two or three factors by phases of three or four experiments. Represented pictorially, these phases are commonly chosen to be equilateral triangles or regular tetrahedra, with a mid-point representing the existing process. These techniques are called Simplex EVOPS,[4] and are illustrated by Figure 9.5.

The procedure is as follows:

(a) Carry out the experiments of the first phase, and note the poorest result within it.

(b) The next phase is then just *one* experiment, the treatment combination for which is the mirror image through the triangle side or tetrahedron face which links the other results.

(c) Succeeding phases also each consist of only one experiment; unless a phase produces two equally poor results, in which case the mirror image of each is

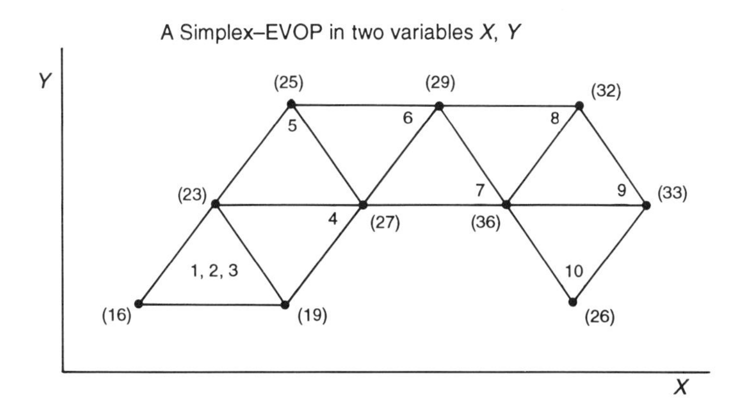

Figure 9.5 *A Simplex-EVOP in two variables X,Y*

tested, and a new phase constructed from the better reflection.

The numbers within Figure 9.5 show the order of experimentation (without brackets) and the results (bracketed). From them, the search would be terminated after point 10, with the assumption that the optimum lies near point 7.

9.2.3 Multi-variable Simplex EVOPS

The advantage of the Simplex method over factorial designs, and the Box EVOPS which were derived from them, is that the total number of treatment combinations in a Simplex procedure does not increase rapidly with the number of variables. Hence, all factors which are suspected of having a bearing on the result should be included in the optimization, since doing so does not greatly increase the size of the investigation. On the other hand, optimization methods in general provide information about only some part (maybe a very small part) of the possible range of responses, since the whole aim is to optimize the response with the least effort. One cannot have it both ways – one cannot study the behaviour of the system thoroughly unless more elaborate experiments, and more detailed analysis, are carried out. As the pioneering text[3] said, '. . . Simplex EVOP can be expected to work well when empirical feedback is the only consideration, but not when, as is usually the case, scientific feedback is of major importance'.

One possible alternative to the factorial and fractional experiments described in Chapters 5–7 is to locate the optimum in a multi-variable system by the Simplex method, and then, in a sense, return to the classical research approach. That is to say, one can measure the effect on the response when one factor is varied while the others are held at their optimum levels, and this procedure can be carried out for each factor in turn. This technique checks that the optimum has been located accurately. In addition, it indicates how important deviations from optimum level are for each factor; sharp changes of response indicate factors which must be carefully controlled. One then has an alternative to the method by which factorial experiments identify significant factors.

9.2.4 The Design of Multi-variable Simplex EVOPS

The development of a Simplex EVOP clearly cannot be done graphically if more than two factors are involved. However, a tabular method is available, as shown for seven factors in Example 9.2a.

Example 9.2a

Since the effects of seven factors are to be examined, the initial phase of the Simplex investigation has eight treatment combinations (known, by analogy with geometrical figures, as vertices). These TCs are carried out, and the responses measured. The procedure to develop the further vertices is shown in Table 9.1.

Table 9.1 *Development of new vertices in multi-variable Simplex EVOP*

	Variables							
Vertex	*A*	*B*	*C*	*D*	*E*	*F*	*G*	*Response*
1	4	15	100	50	0.5	1600	75	80.1
2	6	15	100	80	0.5	1200	55	81.8
3	4	45	100	80	0.3	1600	55	79.2
4	6	45	100	50	0.3	1200	75	76.9
5	4	15	122	80	0.3	1200	75	79.6
6	6	15	122	50	0.3	1600	55	79.8
7	4	45	122	50	0.5	1200	55	74.0
8	6	45	122	80	0.5	1600	75	81.4

The vertex (7) giving the poorest response is rejected, and the factor levels which represent its mirror image are calculated by continuing the Table as follows:

	A	*B*	*C*	*D*	*E*	*F*	*G*
(i) Sum of levels (excluding rejected)	36	195	766	470	2.7	10 000	465
(ii) Centroid of retained levels = sum/n*	5.14	27.9	109	67.1	0.386	1430	66.4
(iii) Rejected vertex	4	45	122	50	0.5	1200	55
(iv) Displacement = (ii)–(iii)	1.14	−17.1	−13	17.1	−0.114	230	11.4
(v) New vertex (9) = (ii) + (iv)	6.28	10.8	96	84.2	0.272	1600	77.8

*where n = number of remaining points.

If the response at vertex 9 was 78.0, vertex 4 would now be deleted. The calculation of vertex 10 would be calculated as in Table 9.1, the experiment performed and the response obtained. The procedure would be repeated until no further improvement was obtained. For practice, show that the levels of vertex 10 are

A	B	C	D	E	F	G
4.36	3.8	117.8	94	0.464	1788	58.7

Since each new vertex is defined by the preceding results, this approach is sometimes known as *self-directing optimization, SDO*; another name is *rotating-Simplex optimization*. The procedure can obviously be extended to any number of factors.

The eight vertices in Table 9.1 do not correspond to a 2^{7-4} fractional factorial, to illustrate the point that there is no need for the first set of TCs (*the starting Simplex*) to be an obviously symmetrical figure in n dimensions. They can be anything which previous experience suggests. But if experience is lacking, one has little guide in deciding how to combine the factor levels for a given TC. One may then wish to use the systematic approach of a fractional factorial.

Example 9.1b

The 2^{5-2} fractional factorial for $A \rightarrow E$, with ACD and BCE as defining contrasts, has TCs (1), *ad, be, abde, cde, ace, bcd, abc*. We know the aliases which may affect our results; they include $A = -CD$, $B = -CE$, $C = -AD$, $-BE$, $D = -AC$, $E = -BC$. If we do the eight TCs, we can (i) analyse for effects, and (ii) delete the two lowest responses to reduce our design to six vertices for five variables. Having done (ii), we then delete the next-lowest response and use this deletion to calculate a new vertex. This procedure is illustrated below. [We could of course use (i) to develop a steepest ascent.] Suppose the results are as in Table 9.2.

Vertices 3, 4 are the initial rejects. Vertex 6 is then rejected and vertex 9 calculated from it: vertex 9 = A 12, B 8.4, C 340, D 5.4, E 2.2.

Table 9.2 *A Simplex in five variables*

Vertex	*TC*	*A*	*B*	*C*	*D*	*E*	*Response*
1	(1)	15	6	300	3	4	47
2	*ad*	30	6	300	5	4	25
3	*bc*	15	9	300	3	7	17
4	*abde*	30	9	300	5	7	10
5	*cde*	15	6	500	5	7	40
6	*ace*	30	6	500	3	7	22
7	*bcd*	15	9	500	5	4	26
8	*abc*	30	9	500	3	4	34

Exercise 9.2

If the response at vertex 9 was 30, what would you do next?

Answer

Delete vertices 2 and 7, since both gave responses <30, and calculate vertices 10 and 11 by deleting 2 and 7 respectively. The new vertices are:

	A	*B*	*C*	*D*	*E*
10	4.8	9.36	556	3.56	4.48
11	25.8	5.16	276	3.56	4.0

9.2.5 Phase Size in Simplex EVOPS

We have seen that factor levels in the starting Simplex depend on our judgement. We are trying to answer the question 'How quickly do we expect the response to change as we change the factor levels: how big should the steps between a given vertex and the previous ones be?' This boils down to 'How big should the starting Simplex be?' As with steepest ascent, the general rule is: change the scales for the various factors so that unit change of level in each factor gives roughly the same change of response.

Unit change here means from low to high in a 2^n experiment, or from low to medium, or medium to high, in a 3^n experiment. Yarbro and Deming[5] showed that the size of the starting Simplex is not critical if the later phases can be contracted or expanded at will; this expansion is brought about very simply by multiplying the displacement values [(iv) in Table 9.1] by >1 or <1 in generating new vertices.

9.2.6 Advantages and Disadvantages of EVOP

Box's original intention for EVOP, that it should gradually optimize manufacturing processes while continuing to produce saleable product, means that it is intended for circumstances when many production runs are intended anyway (which in turn means that the programme itself causes little additional cost). These circumstances are therefore very different from those in the laboratory or pilot plant, where additional time and money must be used to make the runs (but many factors can be handled by skilled staff, and any unusable product is not a major problem).

The obvious advantages of EVOP are that it can improve both process efficiency and product quality, and can also increase the sense of involvement of the process operators. The main disadvantages are that time and money must be spent on training personnel to make the process changes and to keep and analyse simple records. Box's recommendation was that the records be displayed on an information board, to guide the process superintendent and an EVOP committee toward better conditions.

Lowe[6] reviewed these styles of EVOP, and Hahn and Dershowitz[7] reviewed their actual usage. They found that Box EVOP was more used than Simplex, in spite of the greater economy of the latter, and that usage tended to be on short-term pilot plant operations in spite of Box's intentions.

REFERENCES

1. G.E.P. Box and K.B. Wilson, *J. Roy. Stat. Soc., Series B*, 1951, **13**, 1–45.
2. G.E.P. Box, *Appl. Stat.*, 1957, **6**, 81–101.

3. G.E.P. Box and N.R. Draper, 'Evolutionary Operations', Wiley, New York, 1969.
4. W. Spendley, G.R. Hext, and N.R. Himsworth, *Technometrics*, 1962, **4**, 441–461.
5. L.A. Yarbro and S.N. Deming, *Anal. Chim. Acta*, 1974, **73**, 391–398.
6. C.W. Lowe, *Appl. Stat.*, 1974, **23**, 218–226.
7. G.J. Hahn and A.F. Dershowitz, *Appl. Stat.*, 1974, **23**, 214–218.

CHAPTER 10

Factorial Experiments at Three Levels

If a systematic, sequential approach to experimental design is adopted, it is usually sufficient to restrict the design to two-level experiments, supplemented by a five-level 'star' near the end of an investigation. Sometimes, however, it is advantageous to work at three levels. The usual reason for employing either of these techniques is that a maximum or minimum is being approached. As with 2^n experiments, 3^n and higher-level designs indicate maxima and minima, but in addition they estimate the non-linear (quadratic) effects which cause such a circumstance. Three-level experiments were discussed in detail in, *e.g.*, 'Design and Analysis of Industrial Experiments', ed. O.L. Davies, Chapter 8, and in less detail in numerous other texts.[1–3] If the extreme levels are placed symmetrically round the centre one for each factor, the data can conveniently be analysed by a modified Yates table.

10.1 LEVEL SYMBOLS AND DEGREES OF FREEDOM IN 3^n EXPERIMENTS

Consider, *e.g.*, a 3^3 experiment in A, B, C. There will be 27 treatment combinations, and correspondingly 26 degrees of freedom: see Table 10.1.

If the design is duplicated, there will be 53 degrees of freedom. The degrees of

Table 10.1 *Degrees of freedom in a 3^3 experiment*

Effect	ϕ	
I		
A	2	(because three levels of A)
B	2	
AB	4	$(=2\times 2$, from three levels of each of A, $B)$
C	2	
AC	4	
BC	4	
ABC	8	$(=2\times 2\times 2)$
	—	
	26	

freedom for the effects will be the same; hence the error degrees of freedom will be $53-26=27$. The two degrees of freedom for any factor are divided, one for its linear effect and one for its quadratic effect.

For level symbols, use (1), a_1, a_2. . .. The order for Yates analysis will then (corresponding to that for 2^n experiments) be:

$$(1), a_1, a_2, b_1, a_1b_1, a_2b_1, b_2, a_1b_2, a_2b_2, c_1, a_1c_1, a_2c_1, b_1c_1, a_1b_1c_1, a_2b_1c_1, b_2c_1, a_1b_2c_1, c_2, a_1c_2, a_2c_2, b_1c_2, a_1b_1c_2, a_2b_1c_2, b_2c_2, a_1b_2c_2, a_2b_2c_2, \ldots$$

10.2 ANALYSIS BY YATES TABLE

The observations in standard order are divided into sets of three, as shown in Table 10.2. For any set of three responses r_x, r_y, r_z, calculate

(i) the sum $r_x+r_y+r_z$
(ii) the difference r_z-r_x and
(iii) the quantity $(r_x+r_z-2r_y)$

As an example, for the treatment combinations (1), a_1, a_2,

(i) $(1)+a_1+a_2$
(ii) $a_2-(1)$
(iii) $(1)+a_2-2a_1$

A three-level experiment, analysed this way, provides estimates of the linear and non-linear elements of effects as indicated by Figure 10.1

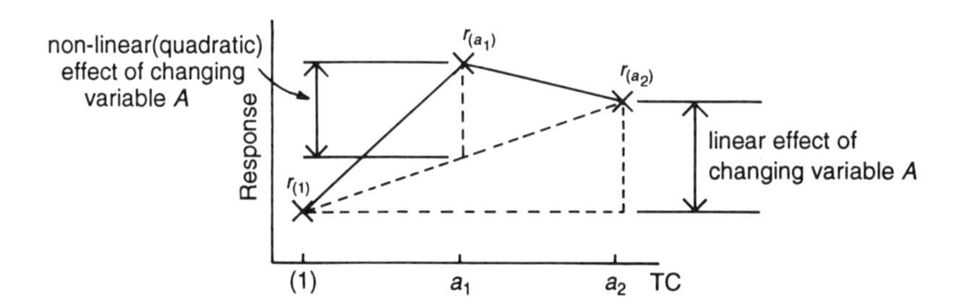

Figure 10.1 *Estimates of linear and quadratic effects from equi-spaced three-level experiments*

If there is no non-linear effect, then, because a_1 is midway between (1) and a_2, $r_{a_1}=\frac{1}{2}(r_1+r_{a_2})$ and $r_1+r_{a_2}-2r_{a_1}=0$. Hence the simple test for non-linearity using equally spaced levels and the expression $r_x+r_z-2r_y$. Note that a *negative* sign for $r_x+r_z-2r_y$ implies an *upward* curve between (1) and a_2 [*i.e.* $r_y>\frac{1}{2}(r_x+r_z)$]. A *positive* sign implies a *downward* curve between (1) and $a_2[r_y<\frac{1}{2}(r_x+r_z)]$.

To calculate the sum of squares for each effect, divide the square of the number in the last column of analysis by $2^m.3^{n-p}.r$, where

m = number of factors in the effect considered
n = number of factors in the experiment
p = number of linear terms in the effect considered
r = degree of replication.

Table 10.2 *Yates analysis for a* 3^2 *experiment with no replication*

TC and response	*Columns of Analysis* 1	2	*Effect*
(1)	$(1)+a_1+a_2$	$(1)+a_1+a_2+b_1+a_1b_1+a_2b_1+b_2+a_1b_2+a_2b_2$	Total
a_1	$b_1+a_1b_1+a_2b_1$	$a_2-(1)+a_2b_1-b_1+a_2b_2-b_2$	A_L
a_2	$b_2+a_1b_2+a_2b_2$	$a_2b_2+b_2-2a_1b_2+b_1-2a_1b_1+(1)+a_2-2a_1$	A_Q
b_1	$a_2-(1)$	$b_2+a_1b_2+a_2b_2-((1)+a_1+a_2)$	B_L
a_1b_1	$a_2b_1-b_1$	$a_2b_2-b_2-(a_2-(1))$	A_LB_L
a_2b_1	$a_2b_2-b_2$	$a_2b_2+b_2-2a_1b_2-((1)+a_2-2a_1)$	A_QB_L
b_2	$(1)+a_2-2a_1$	$b_2+a_1b_2+a_2b_2+(1)+a_1+a_2-2(b_1+a_1b_1+a_2b_1)$	B_Q
a_1b_2	$a_2b_1+b_1-2a_1b_1$	$a_2b_2-b_2+a_2-(1)-2(a_2b_1-b_1)$	A_LB_Q
a_2b_2	$a_2b_2+b_2-2a_1b_2$	$a_2b_2+b_2-2a_1b_2+(1)+a_2-2a_1-2(a_2b_1+b_1-2a_1b_1)$	A_QB_Q

For the effects, the suffix L indicates the linear effect of the factor or interaction. Hence, A_L, B_L, and A_LB_L correspond to the *A*, *B*, and *AB* effects which would arise from a 2^2 experiment. Non-linear (quadratic) effects are indicated by the suffix Q.

Since each effect has one degree of freedom, its sum of squares equals its mean square, as in 2^n experiments. Then, as usual, $MS_{\text{effect}}/MS_{\text{error}}=F_{\text{exptl}}$. The procedure is shown in Example 10.1 for an unreplicated 3^2 experiment to minimize an impurity in a product. We must note that if any variable is *qualitative* (*e.g.* different types of catalyst, different workers, or different sources of raw material), the subdivision into eff_{A_L} and eff_{A_Q} is not appropriate, and we sum the two mean squares and divide by two to give the total *MS* for *A*.

Example 10.1: an unreplicated 3^2 experiment

The TCs $(1)\ldots a_2b_2$, done in random order in an unreplicated experiment, gave responses (impurity concentrations in arbitrary units) as shown in Table 10.3, which completes the calculation to mean squares after deducting 170 from each response.

Table 10.3 *Responses and mean squares in an unreplicated* 3^2 *experiment*

TC	*Response* (*code*: −170)	1	2	$SS=(2)^2/2^m.3^{n-p}=MS$
(1)	−21	−72	0	
a_1	−44	+18	+24	$24^2/6$ = 96
a_2	−7	+54	+144	$144^2/18$ = 1152
b_1	+13	+14	+126	$126^2/6$ = 2646
a_1b_1	−5	−3	−1	$1^2/4$ = 0.25
a_2b_1	+10	+13	−9	$9^2/12$ = 6.8
b_2	+20	+60	−54	$54^2/18$ = 162
a_1b_2	+1	+33	+33	$33^2/12$ = 90.8
a_2b_2	+33	+51	+45	$45^2/36$ = 56.2
	$\Sigma=0$			

Taking the error mean square as the mean of the interaction mean squares gives $MS_{error} = 38.5$. The main factor F-values are then F_{A_L} 2.5, F_{A_Q} 29.9, F_{B_L} 68.7, F_{B_Q} 4.2. Compared with $F_{1,4}$ (single-sided) from F-tables, these show B_L is approaching highly significant ($\alpha \to 0.01$) and A_Q is significant ($0.025 > \alpha > 0.01$).

Example 10.2: a duplicated 3^2 experiment

Now suppose that the responses given in Table 10.3 were the sums of duplicates as in Table 10.4.

Table 10.4 *Responses and mean squares in a duplicated 3^2 experiment*

TC	*Responses (code: −85 from each)*	*Sums*	$\underline{1}$	$\underline{2}$	$SS = (\underline{2})^2/2^m.3^{n-p}.2$	$F_{1,9}$ *(single-sided)*
(1)	−8, −13	−21	−72	0		
a_1	−15, −29	−44	+18	+24	$24^2/12 =$ 48	1.3: NS
a_2	−8, +1	−7	+54	+144	$144^2/36 =$ 576	15.9; 0.01 > α0.001 highly significant
b_1	+2, +11	+13	+14	+126	$126^2/12 =$ 1323	36.4: α < 0.001 very highly significant
a_1b_1	−3, −2	−5	−3	−1	$1^2/8 =$ 0.125	≪1: NS
a_2b_1	+3, +7	+10	+13	−9	$9^2/24 =$ 3.4	<1: NS
b_2	+11, +9	+20	+60	−54	$54^2/36 =$ 81	2.2: NS
a_1b_2	−2, +3	+1	+33	+33	$33^2/24 =$ 45.2	1.3: NS
a_2b_2	+24, +9	+33	+51	+45	$45^2/72 =$ 28.1	<1: NS

The figures in the columns of analysis are unchanged, because they are obtained from the sums of the duplicate results. But the divisor is doubled, since now $r = 2$. Hence the mean squares of the effects are halved. The F-values in Table 10.4 arise from total sum of squares 2432, total treatment sum of squares 2105, error sum of squares 327. Hence the error mean square is 36.3. Comparison with table values for $F_{1,9}$ (single-sided) gives high significance to A_Q and very high significance to B_L. The duplication identifies the interactions as non-significant.

eff_{B_L} is positive; hence to decrease the impurity, decrease the level of B. eff_{A_Q} is positive; hence there is a downward curve from (1) to a_2. Inspection of the data indicates a minimum between these levels. An optimum level for A can now be sought.

Exercise 10.1

A duplicate 3^2 experiment was carried out to estimate the effects of two variables R, S on the active isomer content of a herbicide, with results as follows:

TCs	(1)	r_1	r_2	s_1	r_1s_1	r_2s_1	s_2	r_1s_2	r_2s_2
Run 1	67	60	67	77	72	78	86	83	99
Run 2	62	46	76	86	73	82	84	88	94
Totals	129	106	143	163	145	160	170	171	183

Code the totals by deducting 150 from each, and conduct an ANOVA to show that S_L is very highly significant and R_Q is significant in their effects on the mean.

Answer

$SS = \underline{2}^2/2^m3^{n-p}r = MS$ as follows:

R_L 48, R_Q 300, S_L 1776, R_LS_L negligible, R_QS_L 100, S_Q 32, R_LS_Q 45, R_QS_Q negligible.

Total $SS = 2630$, total treatment $SS = 2301$, error $SS = 329$. Hence error $MS = 329/9 = 36.6$.

$F_{1,9}$ (single-sided) R_L 1.3, R_Q 8.2, S_L 48.5, R_LS_L negligible, R_QS_L 2.73, $S_Q < 1$, R_LS_Q 1.2, R_QS_Q negligible. R_Q $0.025 > \alpha > 0.01$, $S_L \alpha < 0.0001$.

10.3 FRACTIONAL FACTORIALS AT THREE LEVELS

10.3.1 Dividing 3^n Factorials into Blocks

Since each factor has three levels, the number of blocks must be 3^p, where p is less than the number of factors: $p = 1, 2, 3 \ldots$.

Example 10.3: dividing a 3^2 design into three blocks

A 3^2 would be divided into three blocks if (*e.g.*) two variables A, B are to be examined where three batches of raw material (each sufficient for three experiments) are available, or where the total work is to be divided between three workers or three plants. Suppose the three batches (workers, plants...) are identified as α, β, γ. Let the levels of the variables be A_0, A_1, A_2 and B_0, B_1, B_2. The only permissible ways to divide into blocks are those in which α, β, γ each appear once in each row and once in each column (Latin Square designs). Two permissible ways to divide a 3^2 design into three blocks α, β, γ are given in Table 10.5.

In these, α, β, and γ are eliminated when calculating the differences between the means of the rows or columns (*i.e.* calculating eff_A and eff_B respectively). As a result,

Table 10.5 *Two Latin Square designs*

(i)	B_0	B_1	B_2	(ii)	B_0	B_1	B_2
A_0	α	γ	β	A_0	α	β	γ
A_1	β	α	γ	A_1	β	γ	α
A_2	γ	β	α	A_2	γ	α	β

Table 10.6 *Blocks in Latin Square design*

(i)	*TCs*			(ii)	*TCs*		
α	A_0B_0	A_1B_1	A_2B_2	α	A_0B_0	A_1B_2	A_2B_1
β	A_1B_0	A_2B_1	A_0B_2	β	A_0B_1	A_1B_0	A_2B_2
γ	A_2B_0	A_0B_1	A_1B_2	γ	A_0B_2	A_1B_1	A_2B_0

eff_A and eff_B are not confounded with the difference between batches (workers, plants...), assuming, as always in Latin Square and similar designs, that there is no interaction between A, B and α, β, γ.

Rearranging Table 10.5 (i) and (ii) gives blocks α, β, γ in Tables 10.6 (i) and (ii) respectively.

As the design has been divided into three blocks, two effects must be confounded with the between-blocks difference. Since A and B are not confounded, the confounded effects must be two of the four interaction effects, one pair in one design and the other in the other. It can be shown that Table 10.6 (i) confounds A_LB_Q and A_QB_L, and Table 10.6 (ii) confounds A_QB_Q and A_LB_L.

Since a 3^2 design has nine treatment combinations, it has eight degrees of freedom: $\phi_{total}=8$. Each main factor has $\phi=2$, ($\phi_L=1$, $\phi_Q=1$) and the total AB interaction has $\phi=4$ ($\phi A_LB_L=1$, $\phi A_LB_Q=1$, $\phi A_QB_L=1$, $\phi A_QB_Q=1$). The $\phi=2$ for the confounded interaction $=\phi_{error}$. A 3^2 design in three blocks therefore has only two degrees of freedom for error if an estimate of eff_{AB} is to be made and if there is no replication. If it is assumed that eff_{AB} is negligible or zero, then eff_A and eff_B may be compared against the mean of the mean squares for the four AB components.

Example 10.4: dividing a 3^3 design into three blocks

Dividing a 3^2 design into three blocks, confounding elements of the interaction AB, gave the three blocks α, β, γ of Tables 10.6 (i) and (ii). If we now wish to divide a 3^3 design into three blocks, confounding ABC, we must associate the levels of C (C_0, C_1, C_2) with α, β, γ respectively. What confounded elements of AB before will confound elements of ABC now, and the principal block from Table 10.6 (i) is as in Table 10.7 (i).

You will recognize that $A_0B_0C_0$ is the treatment combination usually referred to as (1). It is convenient to simplify the symbols for treatment combinations, compared with Table 10.7 (i) above, by omitting the factor letters but always writing a treatment combination in alphabetical order. Thus, *e.g.*, $A_1B_1C_0$ becomes 110, and the principal block is as in Table 10.7 (ii).

Table 10.7 *Principal block of 3^{3-1} design with $A_LB_QC_Q$ and $A_QB_LC_L$ as defining contrasts*

(i)				(ii)			
	$A_0B_0C_0$	$A_1B_1C_0$	$A_2B_2C_0$		000	110	220
	$A_1B_0C_1$	$A_2B_1C_1$	$A_0B_2C_1$		101	211	021
	$A_2B_0C_2$	$A_0B_1C_2$	$A_1B_2C_2$		202	012	122

Table 10.8 *3^{3-1} Factorial design, with AB^2C^2, A^2BC as defining contrasts*

Principal block	*Block II*	*Block III*
000	100	200
101	201	001
202	002	102
012	112	212
110	210	010
211	011	111
021	121	221
122	222	022
220	020	120

The defining contrasts yielding this principal block are $A_L B_Q C_Q$, $A_Q B_L C_L$.[4] Since the suffixes L and Q denote first- and second-order effects, these symbols translate into $A^1B^2C^2$, $A^2B^1C^1$. Corresponding to the system in 2^n designs in which $A.A = A^2 = 1 (= A^0)$, in 3^n designs, $A^1.A^2 = A^3 = 1 (= A^0)$. The generalized interaction in this case is therefore $AB^2C^2.A^2BC = 1$ (written as I). The product of any set of confounded interactions is always unity; hence $AB^2C^2.A^2BC.I = 1$.

The system utilizing power numbers generates additional blocks from the principal block, as was done for 2^n designs, and the present example develops as below. The remaining two blocks are derived from the principal block by adding one level of one factor to each treatment combination, remembering that if any power number would thereby become 3, it in fact reverts to zero. (This is known as calculating to modulo 3; in the 2^n case, calculation is to modulo 2). The second and third blocks can be obtained by multiplying twice by A to give Table 10.8. Multiplying by B simply transposes blocks II and III.

There are actually four pairs of ABC components which may be used as defining contrasts (AB^2C^2, A^2BC; AB^2C, A^2BC^2; ABC^2, A^2B^2C; ABC, $A^2B^2C^2$, which Yates called W, X, Y, Z interactions respectively). Of these, I used the first in Example 10.4. Any one pair generates three blocks. So four pairs must generate twelve blocks (twelve 3 × 3 Latin Squares), each of which is a one-third replicate of a 3^3 factorial.

The need for equal spacing between the three levels of any factor, for purposes of calculation, need not require linearity in the experimentally set levels. For instance, reagent concentrations 0.5, 1.0, 2.0 M become linear if the calculations are done with log(concentration).

10.3.2 Aliases in 3^n Designs

As in 2^n designs, the aliases of an effect are obtained by multiplying that effect with each of the confounded interactions. In Example 10.4, where those interactions are $I = AB^2C^2 = A^2BC$, it follows that if we multiply by A, we get $A = A^2B^2C^2 = BC$ (the last effect arising from $A.A^2BC = A^3BC = BC$). The full alias matrix is obtained by multiplying the TCs of the full factorial in the basic factors by the confounded interactions (exactly as for 2^n designs). Since the 3^{3-1} design in A, B, C was obtained by adding C to the 3^2 design in A, B, we can regard A and B as the basic factors. The matrix is therefore as in Table 10.9.

Inspection will show that the three columns of the table include all twenty-seven treatment combinations (and corresponding effects) of the 3^3 design. This is to be expected, and is a check of the correctness of the matrix.

If the principal block is done (in random order, of course) and the treatment combinations are put in standard order, Yates table analysis estimates the effects as in Table 10.10. Summarizing from Table 10.9 or 10.10, we find that this design confounds the main factors with two-factor interactions as follows: $A_L = B_L C_L$, $A_Q = B_Q C_Q$, $B_L = A_L C_Q$, $B_Q = A_Q C_L$, $C_L = A_L B_Q$, $C_Q = A_Q B_L$.

Each level symbol (a, a^2, ... c^2) appears three times in the TCs and therefore analysis is possible for each main factor component. But two-factor interactions are aliased with main factors and with one another. Hence, Latin Square and related designs should be used only when interactions are negligible; this point has been made earlier. Only two degrees of freedom are left over for error: Table 10.10

Table 10.9 *Alias matrix for the 3^{3-1} design with AB^2C^2, A^2BC as defining contrasts*

	AB^2C^2	A^2BC
(1)	AB^2C^2	A^2BC
A	$A^2B^2C^2$	BC
A^2	B^2C^2	ABC
B	AC^2	A^2B^2C
AB	A^2C^2	B^2C
A^2B	C^2	AB^2C
B^2	ABC^2	A^2C
AB^2	A^2BC^2	C
A^2B^2	BC^2	AC

Table 10.10 *Yates analyis for the 3^{3-1} design with AB^2C^2, A^2BC as defining contrasts*

TCs	*Effect*	*Two-factor aliases of main factor components*
(1)		
ab	eff_A	BC
a^2b^2	eff_{A^2}	B^2C^2
ac	eff_B	AC^2
a^2bc	eff_{AB}	
b^2c	$eff_{A^2B} = eff_{C^2}$	A^2B
a^2c^2	eff_{B^2}	A^2C
bc^2	$eff_{AB^2} = eff_C$	AB^2
ab^2c^2	$eff_{A^2B^2}$	

shows that for this Example they correspond to eff_{AB} and $eff_{A^2B^2}$. Since main factors are aliased with two-factor interactions, this is only a resolution III design, 3^{3-1}_{III}.

10.3.3 Fractional Three-level Designs in More Than Three Variables

If an experimental run of twenty-seven treatment combinations can be contemplated, then a one-third replicate of a 3^4 design can estimate all main factors clear of two-factor interactions. This is because we can choose, *e.g.*, $AB^2C^2D^2$, A^2BCD as defining contrasts (with, therefore, I as their generalized interaction). As a result, all main factors are aliased only by three- or four-factor interactions; *e.g.* the aliases of A and A^2 are $A^2B^2C^2D^2$, BCD; $B^2C^2D^2$, $ABCD$ respectively. Such a design may be used to estimate the main effects and the two-factor interactions between three of the factors, provided these three do not interact with the fourth. There is a summary and discussion of a variety of multi-factor designs in the text book edited by O.L. Davies, Chapter 10. It can be shown that the defining contrasts $AB^2C^2D^2$, A^2BCD give the 3^{4-1}_{IV} principal block shown as Table 10.87 in that book, and reproduced here as Table 10.11.

Table 10.11 *Principal block of 3^{4-1} design with $AB^2C^2D^2$, A^2BC as defining contrasts*

0000	1010	2020	0120	1100	2110	0210	1220	2200
0021	1001	2011	0111	1121	2101	0201	1211	2211
0012	1022	2002	0102	1112	2122	0222	1202	2212

This principal block arises from multiplying the principal block, block II, and block III of Table 10.7 by D^0, D^1, D^2 respectively. This corresponds exactly to multiplying blocks α, β, γ of Table 10.6 (i) by C^0, C^1, C^2 respectively to form the 3^{3-1} design of Table 10.7. The three designs have AB^2, A^2B; AB^2C^2, A^2BC; $AB^2C^2D^2$, A^2BCD as defining contrasts. If it became necessary to generate a 3^{5-1} design by adding E^0, E^1, E^2 respectively to the principal block of Table 10.11 and its associated blocks II and III, it is logical to expect that the defining contrasts will be $AB^2C^2D^2E^2$, A^2BCDE: a resolution V design is derived.

10.4 LIMITATIONS OF 3^n DESIGNS

In general, fractional designs are not as satisfactory in 3^n design as in 2^n design; relatively large experiments are required to separate first-order interactions from one another and from main factors. Usually the rule is that 2^n fractions are used when the factors are quantitative. We shall see in Chapter 11 how more economical designs, for the estimation of linear, quadratic, and interaction effects, are obtained by combining 2^n factorials or 2^{n-p} fractions with 'star' arrangements of TCs in *central composite designs*. A 2^{5-1} factorial supplemented by a star need take only thirty-six runs (see Table 11.3), compared with the eighty-one TCs of a 3^{5-1} design, maintains resolution V, and has certain other advantages also.

REFERENCES

1. W.E. Duckworth, 'Statistical Techniques in Technological Research', Methuen, London, 1968, Chapter 10.
2. D.C. Montgomery, 'Design and Analysis of Experiments', 2nd Edn., Wiley, New York, 1984, pp. 281–292, 311–317, 349–354.
3. K.C. Peng, 'The Design and Analysis of Scientific Experiments', Addison-Wesley, Reading, Mass., 1967, 126–139.
4. 'The Design and Analysis of Industrial Experiments', ed. O.L. Davies, 4th Edn., Longman, London, 1978, Table 9.62, column W, and p. 426.

CHAPTER 11

Sequential Operations, Composite Designs, and Response Surface Methodology

Response Surface Methodology, RSM, brings together factorial designs and the method of steepest ascent with composite designs (which are discussed later in this chapter) to analyse the kinds of problems with which we are concerned in this book: problems in which independent variables A, B,... influence one or more experimental responses. Common goals of such operations are to determine the best operating conditions for a given machine or process; the identification of operating conditions which will mutually satisfy several criteria (say, two specification limits and the smallest possible production cost for a product); or the ability to predict response variation (say, the change in a product) with variation of A, B,... inside or outside the experimental space investigated.

If A, B,... are continuous variables, then, following Gauss's original idea, the observed response Y is some function of the levels of the variables and a random error component:

$$Y=f(A, B, \ldots)+\varepsilon \qquad (11.1)$$

We saw illustrations in Section 5.7 of the fact that we can calculate expected response values, y, from a process model based on our experimental results. The surface represented by $y=f(A, B, \ldots)$ is called a *response surface*. Two possible surfaces for the dependence of y on two variables are shown in Figures 11.1(a) and (b).

If the real system is well modelled by a planar response surface, and the independent variables can be controlled by the experimenter with negligible error, the responses y predicted from the model will be close to the experimentally observed values Y, where Y is related to the levels of A and B by equation 11.2:

$$Y=\beta_0+\beta_A x_A+\beta_B x_B+\varepsilon \qquad (11.2)$$

This is the case when interactions are zero. Figure 11.1(a) is an illustration, being derived from Example 9.1(a) by re-writing variables X, Y as A, B. Figure 11.1(a) also illustrates a fundamental point: the ideal steepest ascent is perpendicular to the response contours. But if there is marked interaction [Figure 11.1(b)], the results may be poor, unless we are lucky. Hence, we should not do many steepest ascent trials before checking results with prediction.

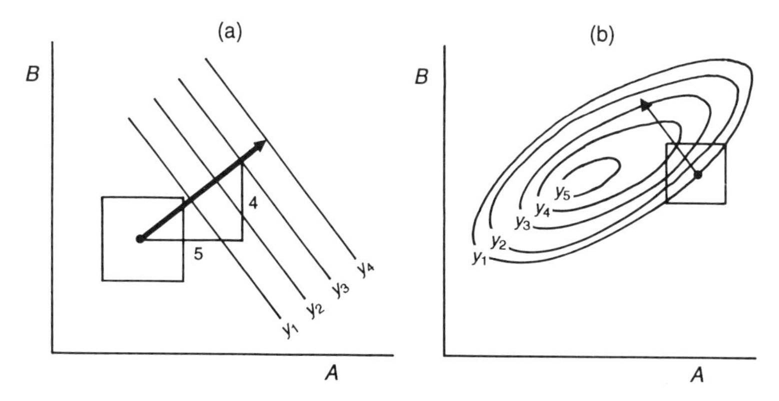

Figure 11.1 *Response surfaces for two variables*

In Equation 11.2, x_A and x_B are the *coded levels* of the corresponding variables. If the actual levels of A in the factorial are A_1, A_2, then the coded levels are

$$\frac{A_1-[0.5(A_1+A_2)]}{0.5(A_2-A_1)}=-1; \quad \frac{A_2-[0.5(A_1+A_2)]}{0.5(A_2-A_1)}=+1$$

For instance, if A_1, A_2 = pH 4, pH 9, then

$$\frac{4-6.5}{2.5}=-1 \qquad \frac{9-6.5}{2.5}=+1$$

The effect of changing pH, eff_A, is estimated across the width of the 2^2 design – *i.e.* across two coded units – and the slope of response Y against coded A is therefore $eff_A/2$. Similarly the slope of response against B is $eff_B/2$.

In Equation (11.2), therefore, β_A and β_B are half the mean effects of A and B as calculated from a 2^2 factorial. Equation (11.2) thus becomes

$$Y=\beta_0+(0.5eff_A)x_A+(0.5eff_B)x_B+\varepsilon \tag{11.3}$$

Equations (11.2) and (11.3) show that β_0 has the value $Y_0-\varepsilon_0$, corresponding to the situation when the levels of A and B are zero (*i.e.* at the mean of the experimental levels -1, $+1$, and therefore at the mean of the 2^2 design. β_0 is the experimental mean, and ε_0 is its error (its deviation) from the true but unknown mean Y_0.

A 2^2 factorial (in which $eff_{AB}=0$) gives estimates of the mean, eff_A, and eff_B, and therefore gives the *response surface equation*:

$$y=\text{mean}+(0.5eff_A)x_A+(0.5eff_B)x_B \tag{11.4}$$

which can be used to predict values of y at levels x_A, x_B other than -1, $+1$. If we re-write Example 9.1(a) to give the following results

TC	(1)	*a*	*b*	*ab*
Response	30	35	34	39

we can substitute in equation (11.4) as follows:

TC	$\text{Response} = \text{mean} + (0.5eff_A)x_A + (0.5eff_B)x_B$
(1)	$30 = 34.5 - 0.5eff_A - 0.5eff_B$
a	$35 = 34.5 + 0.5eff_A - 0.5eff_B$
b	$34 = 34.5 - 0.5eff_A + 0.5eff_B$
ab	$39 = 34.5 + 0.5eff_A + 0.5eff_B$

so

$$a - (1) = ab - b = eff_A = +5$$
$$b - (1) = ab - a = eff_B = +4$$

[exactly as in Example 9.1(a)].

If there is an interaction, the equation becomes

$$y = \text{mean} + (0.5eff_A)x_A + (0.5eff_B)x_B + (0.5eff_{AB})x_A x_B \qquad (11.5)$$

where x_A, x_B each again have values -1, $+1$ and they multiply algebraically.

In contrast to equations (11.2)–(11.4), equation (11.5) is *second-order in the variables* since it contains the multiplicative term $x_A x_B$.

Exercise 11.1

To show that equation (11.5) is correct, show that it produces the responses of Exercise 9.1 when the mean response and half the estimates of eff_D, eff_E, and eff_{DE} from that exercise are substituted in equation (11.5).

Answer

TC	$\text{Response } y = \text{mean} + (0.5eff_D)x_D + (0.5eff_E)x_E + (0.5eff_{DE})x_D x_E$ (11.6)
(1)	$297 = 201.5 - (-60.5) - (-25.0) + (+10)$
d	$156 = 201.5 + (-60.5) - (-25.0) - (+10)$
e	$227 = 201.5 - (-60.5) + (-25.0) - (+10)$
de	$126 = 201.5 + (-60.5) + (-25.0) + (+10)$

Equations such as (11.3)–(11.5) are *regression equations* which summarize how the response varies with the coded levels and effect magnitudes of the factors (and interactions, where these occur). The halved mean effects are the *regression coefficients* of the equations. These equations are used to derive response contours, and therefore response surfaces, as follows.

11.1 DERIVATION OF RESPONSE SURFACES

The way to do this is to rearrange, say, equation (11.5) so that different values of x_A can be calculated for a fixed value of y and a sequence of values of x_B.

Example 11.1

The expected breaking strain y from Exercise 9.1 is given by equation (11.6). From this,

$$x_D = \frac{\text{mean} + (0.5\mathit{eff}_E)x_E - y}{-0.5\mathit{eff}_D - (0.5\mathit{eff}_{DE})x_E} \tag{11.7}$$

Now, $\mathit{eff}_D = -121$, $\mathit{eff}_E = -50$, $\mathit{eff}_{DE} = +20$, mean $= 201.5$, and therefore

$$x_D = \frac{201.5 - 25x_E - y}{60.5 - 10x_E} \tag{11.7a}$$

The contour corresponding to $y = 200$ will therefore include x_D, x_E coded and uncoded values such as:

$$\text{when } x_E = -1,\ x_D = \frac{201.5 + 25 - 20}{60.5 + 10} = 0.38 \text{ (coded)}$$
$$\equiv 95 + 0.38(5)$$
$$= 96.6 \text{ (uncoded)}$$

$$\text{when } x_E = 0,\ x_D = \frac{201.5 - 200}{60.5} = 0.02 \text{ (coded)}$$
$$\equiv 95 + 0.02(5)$$
$$= 95.1 \text{ (uncoded)}$$

$$\text{when } x_E = +1,\ x_D = \frac{201.5 - 25 - 200}{60.5 - 10} = -0.47 \text{ (coded)}$$
$$\equiv 95 - 0.47(5)$$
$$= 92.6 \text{ (uncoded)}$$

Iteration by computer of this procedure allows a table of uncoded x_D, x_E values to be derived for a range of y values, providing a corresponding range of contours. When the values of x_D so derived become too far from the region of interest, the equation for x_E may be used instead. In this example, the equation is

$$x_E = \frac{\text{mean} + (0.5\mathit{eff}_D)x_D - y}{-0.5\mathit{eff}_E - (0.5\mathit{eff}_{DE})x_D} = \frac{201.5 - 60.5x_D - y}{25 - 10x_D} \tag{11.8}$$

Exercise 11.2

Suppose a duplicated 2^2 factorial in A, B gave sum responses as follows:

TC	$-1,-1$	$+1,-1$	$-1,+1$	$+1,+1$
Sum	77	53	91	67

Sketch the contours for $y = 20, 30, 40, 50$ over coded $x_B = -2, -1, 0, +1, +2$, and compare with Figure 11.2.

Answer

$$x_A = (36.0 + 3.5x_B - y)/6$$

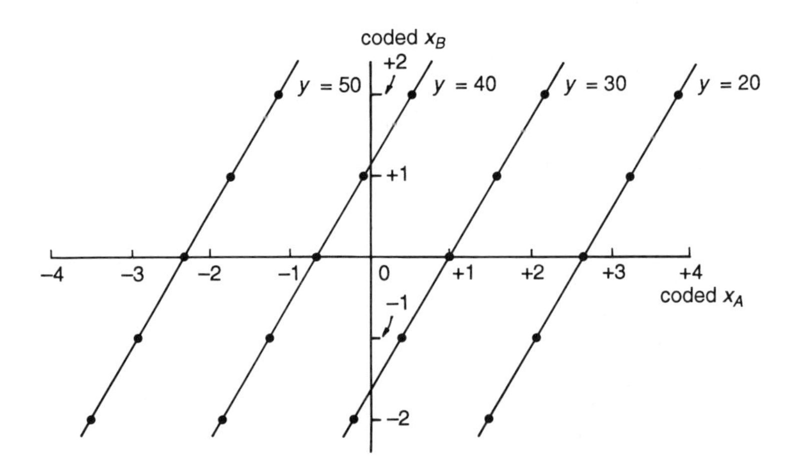

Figure 11.2 *Contours of* $y = 36.0 - 6x_A + 3.5x_B$

11.1.1 Uncertainty of Response Contours

The facts that the lines are sharp in Figures 11.2 and 11.3 does not mean that the fitted values y are exactly known. In fact, it can be shown that the variance of y is at a minimum at the centroid and increases in all directions on moving away from there. For instance, in a 2^2 factorial, the variance is given by

$$var(y) = \frac{\sigma^2}{n}(1 + x_A^2 + x_B^2 + x_A^2 x_B^2) \tag{11.9}$$

in which σ^2/n is the standard error of the mean and x_A, x_B are coded levels.[1] Hence, $var(y) = \sigma^2/n$ at the centroid and, *e.g.*, $4\sigma^2/n$ at the factorial TCs. Uncertainty is thus least at the centroid (we put most confidence in the mean) and any given y contour is increasingly imprecise farther away. Compare the confidence intervals for a regression line (Section 2.5.6 and Figure 2.3). The traditional need for caution in extrapolation outside the experimental space is apparent from these remarks. Also, of course, the response equation should not be expected to hold indefinitely outside that space; and, in addition, systematic errors may lead to bias in the results.[2]

11.2 THE IMPORTANCE OF THE CENTROID

If there is no response surface curvature [as in equation (11.3) and Exercise 11.2], the mean of the responses from the individual TCs will equal the response at the centroid (assuming error is zero). If there is curvature, this equality will not exist, and it becomes important to determine the experimental response at the central point. Carrying out a number of runs at the centroid to determine the mean centroid response, $\bar{y}_c$, has three important applications, as follows:

(a) One check of the appropriateness of the first-order model, such as equation

(11.3), is to compare y_c with the mean $\bar{y}$ of the factorial TCs. $\bar{y}-\bar{y}_c$ is one measure of the overall curvature of the surface, and it can be shown that $\bar{y}-\bar{y}_c$ is an estimate* of the quadratic regression coefficients β_{AA} and β_{BB}:

$$E\{\beta_{AA}+\beta_{BB}\}=\bar{y}-\bar{y}_c \tag{11.10}$$

where E = estimate.

(b) The responses from the replicated centroid can be used to give an estimate of the *'pure' experimental error variance*, to allow F-testing of (i) the estimated effects from an unreplicated 2^n experiment, and (ii) the estimate $\bar{y}-\bar{y}_c$ of the curvature. These procedures are illustrated in Example 11.2.

If the replicates are to give a reliable estimate, they must be scattered randomly throughout the total experiment, and be subject to all the usual errors of setting-up, sampling, synthesis, and analysis; they must not be a special set.

(c) A further, very important reason for replication of the centroid is that the centroid response can have profound effects on the general shape and orientation of the derived response surface. For simplicity, imagine 3^2 factorials are done, giving responses as in Figure 11.3a, b, and c. All the responses but one are the same in each case, but the different centroid responses give markedly different response surfaces. Hence, *of all TCs, the centroid is the one most demanding of replication.*

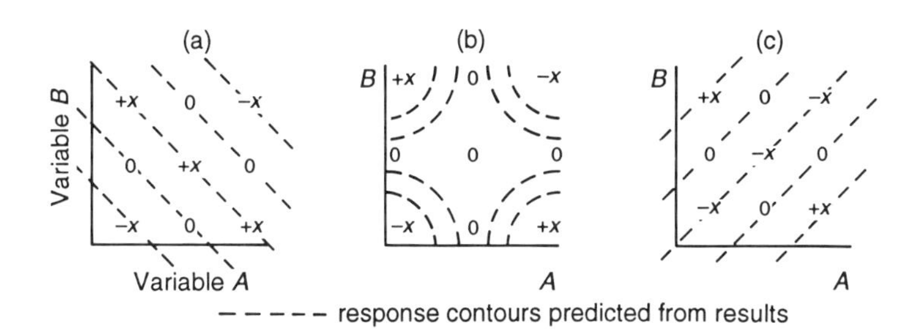

Figure 11.3 *Three 3^3 factorial experiments differing only in centroid response*

11.3 SEQUENTIAL DEVELOPMENT OF AN INVESTIGATION

RSM is a *sequential procedure*. There is often little curvature in the response surface far from the optimum, and a first-order model will be appropriate. If so, our objective is to proceed rapidly and efficiently to near the optimum. In the latter region, a more elaborate surface may be evaluated, such as a second-order model. The principles of the statistical analysis are then complex except for those adept in matrix algebra, and are not explained here. The subject is treated in great detail in the text by Box and Draper, and appropriate software is readily available to carry out such analyses. Example 11.2 below illustrates the earlier stages of the development and Section 11.4 introduces the most widely used design type for

* Except in the rare case of the 'minimax' (Figure 11.3b).

fitting second-order models. Software packages for experimental design, and the kind of information which is easily obtained, are looked at in Appendix 1.

Example 11.2: process-model fitting by first-order equations

Assume that the data from an unreplicated 2^n factorial in J, K are supplemented by five determinations at the centroid, giving results as in Table 11.1a. The aim is to fit a first-order model to give a steepest descent direction, with the supplemental ability to check the adequacy of that model. The purpose of the experiment was to minimize the consumption of raw material per unit amount of wanted product.

Analysis of the 2^2 factorial gives results as follows:

Effects	*Regression coefficients*	*Sums of squares*
J	0.85	2.89
K	0.50	1.00
JK	−0.10	0.04

The repeat observations at the centroid are used to calculate an estimate of the pure experimental error variance, MS_{pe}, as follows:

$$MS_{pe} = \frac{40.3^2 + 40.8^2 + 40.5^2 + 40.6^2 + 40.8^2 - (203.0)^2/5}{4}$$

$$= \frac{8241.98 - 8241.80}{4} = 0.045 \quad (\phi = 4)$$

The difference between the factorial mean, $\bar{y}$, and the centroid mean, $\bar{y}_c$, may be used like this:

$$\bar{y} - \bar{y}_c = -0.05 = \beta_{JJ} + \beta_{KK}$$

where β_{JJ}, β_{KK} are the regression coefficients of the quadratic terms.

It can be shown that the sum of squares (quadratic) is given by

Table 11.1a *Response data from a 2^2 factorial with replicated centroid*

Actual levels		*Coded levels*				
J (*pressure*/atm)	K (T°C)	J	K	*Response* y		
145	180	−1	−1	39.1	$\bar{y} = 40.55$	$\bar{y} = 40.58$
145	200	−1	+1	40.3		
165	180	+1	−1	41.0		
165	200	+1	+1	41.8		
155	190	0	0	40.3	$\bar{y} = 40.60$	
155	190	0	0	40.8		
155	190	0	0	40.5		
155	190	0	0	40.6		
155	190	0	0	40.8		

$$SoS_{\text{quadratic}} = n_1 n_2 (\bar{y} - \bar{y}_c)^2 / (n_1 + n_2) \qquad (11.11)$$

where n_1 = number of TCs in the factorial and n_2 = degree of replication of the centroid; so here $SoS_{\text{quadratic}} = 20(0.05)^2/9 = 0.006$ ($\phi = 1$).

The analysis of variance compares F_e values with F_{tables} for $\phi = 1, 4$ (Table 11.1b).

The interaction and quadratic coefficients are not significant. There is no need to question the adequacy of the first-order model, which we write as:

$$y = 40.58 + 0.85x_J + 0.50x_K$$

Another test of the validity of the model is to compare the regression coefficients (here, 0.85 and 0.50) with their standard error, which for each is $\sqrt{MS_{\text{error}}/n} = 0.07$. The coefficients are large compared with this: they are significantly not zero. The confidence limits for any coefficient b are $b \pm [t_{0.5} \cdot SE_{(b)}]$.

Because there is no evidence of an interaction term, we are probably far from the optimum, and we need to move towards one. From the positive regression coefficients, increasing *J* or *K* increases the consumption of raw material. We must therefore pursue a path of steepest descent in *J* and *K*. Again taking account of the coefficients, we would move 0.85 coded units of *J* down for every 0.5 units of *K* down, starting at $x_J = x_K = 0$. Assume for simplicity that 0.85 coded units of *J* constitutes a convenient step size, and suppose that the responses along the path are as shown in Table 11.1c.

Another 2^2 must now be carried out around *J* = 121 atm, *K* = 170 °C, with replication at the centroid, to develop a new first-order model. Because the

Table 11.1b *Analysis of variance from table 11.1a*

Effect	*SoS*	ϕ	*MS*	F_e	α, *single-sided*
J_L	2.89	1	2.89	64	<0.001, VHS
K_L	1.00	1	1.00	22	<0.01, HS
$J_L K_L$	0.04	1	0.04	<1	NS
$J_Q + K_Q$	0.006	1	0.006	<1	NS
Error	0.18	4	0.045	—	—

Table 11.1c *Steepest descent experiment*

Coded levels		*Actual levels*		
J	*K*	*J*	*K*	*Response y*
0	0	155	190	within original design –
−0.85	−0.5	146.5	184	not normally replicated
−1.7	−1.0	138	180	38.6
−2.55	−1.5	129.5	175	37.7
−3.4	−2.0	121	170	36.7
−4.25	−2.5	112.5	165	38.0

gradients dy/dJ, dy/dK are likely to be smaller nearer the optimum, it is sensible to increase the size of a coded unit. Assume that the new units are 15 atm, 15 °C, and assume that the responses are as in Table 11.1d.

Analysis of the 2^2 factorial gives results as follows:

Effects	*Regression coefficients*	*Sums of squares = mean squares*
J	0.28	0.30
K	0.22	0.20
JK	0.22	0.20

The sum of squares for pure experimental error is given by

$$37.0^2+36.8^2+37.0^2+36.6^2+36.6^2-(184)^2/5=0.16$$

so

$$MS_{\text{pe}}=0.04.$$

The F_{e}-values for J, K, and JK are therefore 7.5, 5.0, and 5.0 ($\phi=1, 4$) giving $\alpha>0.05$, <0.10, <0.10 respectively. At this level of analysis, therefore, eff_J is approaching significance and eff_K and eff_{JK} may possibly be significant. If we stopped there, we would have a model

$$y=37.20+0.28x_J+0.22x_K+0.22x_Jx_K$$

to consider by further experimentation.

But the test of significance for the quadratic effect is

$$\frac{MS_{\text{quadratic}}}{MS_{\text{error}}}=\frac{n_1n_2(\bar{y}-\bar{y}_c)^2/(n_1+n_2)}{0.04}=\frac{20(1.24)^2/9}{0.04}=85 \quad (\phi=1, 4)$$

For this, α(single sided) <0.001: very highly significant. The curvature so indicated means we may be near the optimum. Additional experiments must be done to locate the optimum more precisely.

Table 11.1d *Responses from second 2^2 factorial with replicated centroid*

Actual levels		*Coded levels*				
J	K	J	K	*Response y*		
106	155	−1	−1	37.6	$\bar{y}=37.88$	$\bar{y}=37.20$
136	155	+1	−1	37.7		
106	185	−1	+1	37.6		
136	185	+1	+1	38.6		
121	170	0	0	37.0	$\bar{y}=36.64$	
121	170	0	0	36.8		
121	170	0	0	37.0		
121	170	0	0	36.6		
121	170	0	0	36.6		

Exercise 11.2

Responses from a second-order composite design were as shown in Table 11.2. Analyse the 2^2 data to give an equation of the form

$$y = \text{mean} + b_A x_A + b_B x_B + b_{AB} x_A x_B$$

(As is common practice, use all the seven responses to give the mean in this equation.) Calculate MS_{pe} from the centroid, and calculate $F = MS_{\text{effect}}/MS_{\text{pe}}$ for A, B, AB. Also calculate $MS_{\text{quadratic}}/MS_{\text{pe}}$, and show from these two tests that the equation given above is inappropriate.

Answer

Analysing the factorial data gives $b_A = -1.65$, $b_B = +1.10$, $b_{AB} = -4.00$, and all the Ys give $\bar{Y} = 70.3$. Hence,

$$y = 70.3 - 1.65x_A + 1.10x_B - 4.00x_A x_B$$

From the centroid,

$$MS_{\text{pe}} = \frac{74.0^2 + 71.6^2 + 72.8^2 - (218.4)^2/3}{2} = 1.45$$

The sums of squares (= mean squares) for an unreplicated 2^2 are $2^2 b^2$, so:

	A	B	AB
$F_{1,2} =$	10.89/1.45	4.84/1.45	64.0/1.45
$=$	7.51	3.34	44.1
α	>0.10	>0.10	<0.025
	NS	NS	SIG

The interaction effect is greater than the main effects: we are probably near the optimum.

Additionally,

$$\frac{MS_{\text{quadratic}}}{MS_{\text{pe}}} = \frac{12(68.45 - 72.8)^2/7}{1.45} = 22.4\text{: } 0.05 > \alpha > 0.025$$

The equation is inappropriate. The design should be augmented to form a

Table 11.2 *Responses from a factorial design in variables A, B with replicated centroid*

Actual levels		*Coded levels*		*Response y*
A	*B*	*A*	*B*	
50	90	−1	−1	65.0
80	90	+1	−1	69.7
50	110	−1	+1	75.2
80	110	+1	+1	63.9
65	100	0	0	74.0
65	100	0	0	71.6
65	100	0	0	72.8

composite design, which will allow the fitting of an equation including quadratic terms:

$$y = b_0 + b_A x_A + b_B x_B + b_{AA} x_A^2 + b_{BB} x_B^2 + b_{AB} x_A x_B \quad (11.12)$$

11.4 COMPOSITE DESIGNS

We would like to have as complete as possible a picture of round the optimum, so that we can be confident that:

(a) we are near that optimum (although we should be aware that it may be only a local optimum, with another elsewhere)
(b) if we are not, we have clues as to how to move towards it
(c) we have estimates of the surface curvature: if it is steep, the process will be difficult to control.

An equation such as equation (11.12) may be suitable. In that equation, there are six coefficients to be determined. If the design is in three variables, ten coefficients must be estimated:

$$Y = \beta_0 + \beta_A x_A + \beta_B x_B + \beta_C x_C + \beta_{AA} x_A^2 + \beta_{BB} x_B^2 + \beta_{CC} x_C^2 + \beta_{AB} x_A x_B + \beta_{AC} x_A x_C + \beta_{BC} x_B x_C + \varepsilon \quad (11.13)$$

These βs of course correspond to the mean, the linear, and quadratic effects of A, B, C, and the effects of their interactions in pairs. The model ignores higher-order interactions.

A design for fitting a second-order model must have at least three levels of each factor; but Box *et al.* concluded that 3^n designs are not particularly helpful to experimenters.[3] In developing designs to solve such problems, and to make the information more helpful, they found that the calculations can be simplified if the variance of y is a function only of distance from the centroid, and irrespective of direction (*rotatable designs*). To achieve this and to introduce the extra TCs necessary, Box and Wilson[4] introduced these extra points on the perpendicular axes through the centroid. Such points are called *star points*, and the centroid is added to the design. When added to a 2^n factorial, the whole is called a *central composite design*. Hence, *e.g.*, a design for three variables has six star points, two equidistant along each axis through the centroid.

In general, an unreplicated central composite design consists of a 2^n factorial, or 2^{n-p} fraction, coded to the usual ± 1 notation, augmented by replicated centre points and $2n$ axial points $(\pm\alpha, 0, 0, \ldots)$, $(0, \pm\alpha, 0, \ldots)$, $\ldots$ $(0, 0, \ldots \pm\alpha)$.* The general expression for α, for any 2^{n-p} factorial replicated r_C times, and combined with $2n$ axial points replicated r_s times, is

$$\alpha = (2^{n-p}.r_c/r_s)^{0.25} \quad (11.13)$$

For the simplest case of a full factorial ($p=0$) without replication ($r_c = r_s = 1$), this becomes $\alpha = (2^n)^{0.25}$. The designs for 2^2 and 2^3 are shown in Figure 11.4a and b. These, and numerous other designs, are described in detail in Box and Draper[5]

* This alpha is of course no relation of the probability α used elsewhere in this book.

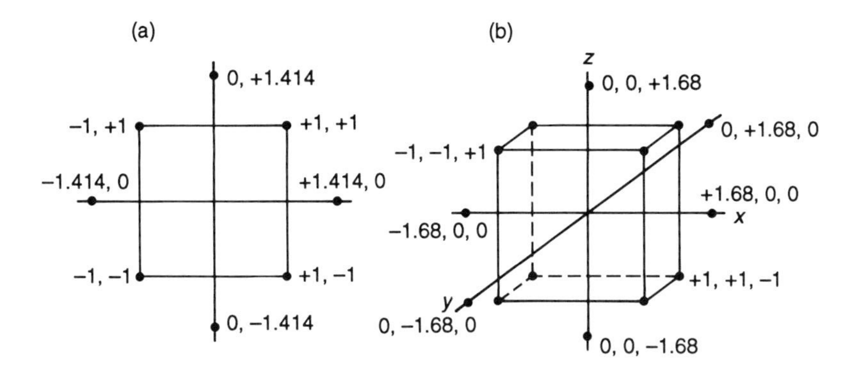

Figure 11.4 *Central composite designs based on 2^2 and 2^3 factorials without replication*

and relatively briefly in other texts.[6–8] If a fractional factorial is used, it is often of resolution V or greater, so that no second-order interaction is aliased with less than a third-order. If so, at least five factors must be involved, using the five-factor interaction to divide the design into two. However, since the star portion yields extra information on the main factors, resolutions as low as III can be used, provided two-factor interactions do not alias one another.[9]

The minimum size for an unreplicated composite design would have only one centroid determination, with total TCs $= 2^n + 2n + 1$; but usually several (say n_0) centre points are done, as explained in Section 11.2. Therefore, for only two factors, a composite design is less economical than a 3^2 design. But from $n = 3$ onwards, composite designs are increasingly economical. Useful designs are to be found in Box and Draper, Table 15.2 and Montgomery, Table 15.8. Some simpler designs are shown in Table 11.3.

The total n_0 depends on several criteria,[10] and the sizes shown in Table 11.3 are convenient compromises. If more centre points are done (for instance, to improve the pure error estimate) nothing is lost except the cost of doing so. The n_0 centre points are divided into n_{c0} factorial centre points and n_{s0} star points. The factorial may be split into blocks, dividing the n_{c0} points equally between blocks. If the star is replicated, there is equal allocation of the n_{s0} points between replications.

The general method of *analysis of composite designs* is first to test the factorial plus n_{c0} data, as demonstrated in Example 11.2 and Exercise 11.2. When the star and n_{s0} responses are added, the whole data are analysed together. Figure 11.4 shows

Table 11.3 *Second-order composite designs*

	Number of factors, n	2	3	4	5	5 (*half replicate*)
Treatment combinations	2^n	4	8	16	32	16
	Axial points	4	6	8	10	10
	Centre points	8	9	12	17	10
		$\alpha = 1.414$	1.682	2.000	2.378	2.000

that this is now a *five-level* design. The data analysis is based in the theory of matrices, and a recent discussion is that by Morgan.[8] However, the matrices are too large to handle except by computer programs, and no description of the theory is attempted here. Some software packages which include appropriate routines are reviewed in Appendix 1.

11.4.1 Effects and Aliases in Composite Designs

Example 11.2 and Exercise 11.2 showed that when interactions are significant in 2^n designs, the observed mean response at the centroid, y_c, is significantly different from the value expected from the 2^n TCs, $\bar{y}$. Matrix theory shows that this is because the quadratic effects of the factors are *co-variates* of the true mean; in the language we have used so far, they are aliases of one another. Hence, for a 2^2 factorial in A, B, the observed mean b_0 is an estimate not of the true mean β_0, but of $\beta_0 + \beta_{AA} + \beta_{BB}$. But we can estimate β_A, β_B, β_{AB} without aliasing (their co-variance is zero: they each have a different effect equation).

The same situation exists in the composite design for A, B. Now, however, we wish to estimate β_{AA} and β_{BB}, and aliasing between them is only zero if the design is such that

$$N n_c = (n_c + 2\alpha_1^2)^2$$

where

$$N = \text{total TCs in the design} = n_c + n_s + n_0$$
$$n_c = \text{factorial TCs}$$

which gives

$$\alpha_1 = \frac{\sqrt{N n_c} - n_c}{2}$$

This α_1 is the multiplier of the coded values ± 1 of the factorial, to give the co-ordinates of star points for minimum aliasing, and is numerically equal to $\alpha = (n_c)^{0.25}$ only for certain values of n_0. To give a rotatable design, we choose α. Then we add to the number of centre points, n_0, until (as closely as possible) $\alpha = \alpha_1$.

For instance, starting with a 2^2 factorial, $\alpha = (4)^{0.25} = 1.414$, and the composite design will include $n_c = 4$, $n_s = 4$, and as many centre points as necessary. To equate $\alpha = \alpha_1$, we write

$$\alpha_1^2 = \frac{\sqrt{n_c(n_c + n_s + n_0)} - n_c}{2} = \alpha^2$$

$$= \frac{\sqrt{4(8 + n_0)} - 4}{2} = \alpha^2$$

giving $n_0 = 8$.

The ideal central composite design in two factors A, B, is therefore as shown in Table 11.4, which explains the entry for $n = 2$ in Table 11.3. Corresponding

Table 11.4 *2^2 factorial design augmented with star design to form a central composite design*

Factorial		*Star*	
A	*B*	*A*	*B*
−1	−1	$-\sqrt{2}$	0
+1	−1	$+\sqrt{2}$	0
−1	+1	0	$-\sqrt{2}$
+1	+1	0	$+\sqrt{2}$
0	0	0	0
0	0	0	0
0	0	0	0
0	0	0	0

(Variables in coded units; TCs done in random order.)

calculations give the designs of Table 11.3 for $n > 2$, which are therefore examples of designs for best rotatability and minimum aliasing.

11.4.2 Lack-of-fit Test for Composite Designs

In Example 11.2 and Exercise 11.2 we saw that using the ratio $MS_{\text{quadratic}}/MS_{\text{error}}$ demonstrated that a model equation without quadratic terms was inadequate. Similarly, in analysing composite designs, the quadratic effects may exceed the first-order effects, suggesting that the new model is becoming inadequate. A new lack-of-fit test is required, and this may be explained as follows.

The total sum of squares of responses Y about their mean $\bar{Y}$ measures the total amount of variability in the results. It can be subdivided into a regression sum of squares and an error sum of squares as follows:

$$\sum_{j=1}^{n} (Y_j - \bar{Y})^2 = \sum_{j=1}^{n} (y_j - \bar{Y})^2 + \sum_{j=1}^{n} (Y_j - y_j)^2 \tag{11.14}$$

$$SoS_{\text{total}} = SoS_{\text{regression}} + SoS_{\text{error}}$$

where the Ys are experimental values and the ys are values calculated from the chosen regression equation, say,

$$y = b_0 + b_A x_A + b_B x_B + b_{AA} x_A^2 + b_{BB} x_B^2 + b_{AB} x_A x_B \tag{11.15}$$

The regression sum of squares is a measure of the variability accounted for by equation (11.15). The error sum of squares is a measure of unaccounted variability. Equation (11.14) shows that it is the sum of squares of the difference between corresponding observed and calculated responses.

The error SoS can be subdivided:

$$SoS_{\text{error}} = SoS_{\text{pure error}} + SoS_{\text{lack of fit}} \tag{11.16}$$

$SoS_{\text{pure error}}$ (henceforth SoS_{pe}) is calculated as shown in Example and Exercise 11.2, and $SoS_{\text{lack of fit}}$ (SoS_{lof}) is calculated by difference.

Suppose a central composite design in two factors is carried out with 2^2 factorial TCs, 2×2 axial points, and five centroid replications: $n_{total} = 13$ and $\phi_{total} = 12$. If the responses are fitted to equation (11.15), there are six coefficients, which must account for (6 − 1) degrees of freedom. [This is equivalent to five effects, $0.5(b_A, b_B \ldots b_{AB})$, each with $\phi = 1$.] Thus $\phi_{error} = 7$. Of these, $\phi_{pe} = 4$, so $\phi_{lof} = 3$. Hence $SoS_{pe}/4 = MS_{pe}$; $SoS_{lof}/3 = MS_{lof}$.

MS_{lof} is a measure of the difference between the experimental mean and the mean predicted from the equation. If this is large compared with the experimental error at that point, as measured by MS_{pe}, the equation is a poor fit to the data and should be rejected. Hence, if $F = MS_{lof}/MS_{pe}$ is significant compared with F_{tables} (here, for $\phi = 3, 4$), the equation should be rejected: *model adequacy* has not been demonstrated.

11.4.3 Other Checks of Model Adequacy

If the F-value from the lack-of-fit test is not significant, this is only one check of adequacy, and caution suggests others should be tried. As in Chapter 5, if the model is adequate, the sizes of the residuals $Y - y$ should have no obvious relationship to sizes of the ys, and should give a reasonable straight line when plotted on normal probability paper. If there are enough of them, their relative frequencies should bear a reasonable resemblance to the Normal distribution.

11.4.4 Moving on from an Inadequate Model

If the model is judged inadequate, it is common to add third-order terms to the polynomial, and to continue until a very good fit is obtained. It is particularly tempting to do this, because computers can quickly give a very good fit to any set of data by adding enough terms. But this may add little to the experimenter's understanding (it adds terms, including higher-order interactions, which are difficult to interpret). It is also an unsure basis for extrapolation.

Instead, it is often advantageous to consider transforming the factors and/or the responses. The chemist's knowledge may suggest that the responses are simply related not to the levels of the factors, but to their logarithms. A logarithmic transformation is also indicated if the residuals show a proportional relationship to the effects (a common defect with many measuring instruments). Some other possible transformations have been summarized in Section 5.7. The aim in any transformation is to minimize the error variance. When a model has been selected, it should be *tested* at points beyond the original design; so-called 'confirmatory runs' at the believed optimum merely check reproducibility at that point.

11.4.5 Analysis of Variance in Composite Designs

Equations (11.14) and (11.16) do not only explain the lack-of-fit test; they are also the basis of the analyses of variance offered by software programs:

(a) an ANOVA for the regression equation and the lack-of-fit and pure error estimates, and

Table 11.5 *Analyses of variance for a composite design in factors, A, B, C*

Source of variation	*SoS*	ϕ	*MS*	F_e
(a) Regression equation	SoS_R	3	$SoS_R/3$	MS_R/MS_r
Residual	SoS_r	8	$SoS_r/8$	—
lack-of-fit	SoS_{lof}	5	$SoS_{lof}/5$	MS_{lof}/MS_{pe}
pure error	SoS_{pe}	3	$SoS_{pe}/3$	—
Total	Sum of above	11		
(b) Regression equation	SoS_R	6	$SoS_R/6$	As above
Residual	SoS_r	5	$SoS_r/5$	
lack-of-fit	SoS_{lof}	2	$SoS_{lof}/2$	
pure error	SoS_{pe}	3	$SoS_{pe}/3$	
Total	Sum of above	11		
(c) Input variable	Regression coefficient / Standard error $= t_e$			

(b) an ANOVA for the individual factors and interactions which we wish to study.

The precise details will depend on the process model we choose.

Suppose we carry out a 2^3 factorial plus four centre points to study three factors *A*, *B*, *C*, which we believe will affect the outcome of our work. If we believe we know enough of the mechanism to say that the result will be zero if we set *A*, *B*, and *C* at zero, we predict $\beta_0 = 0$ and write the predicted *y*s as

$$y = \beta_A x_A + \beta_B x_B + \beta_C x_C$$

The analysis of variance will then be as indicated in Table 11.5(a). On the other hand, if we choose not to assume $\beta_0 = 0$ and we wish to include first-order interactions, the program analyses the data in terms of equation (11.15), and produces an ANOVA as in Table 11.5(b), with the calculation of *F*s as in Table 11.5(a). The programs also produce the regression coefficients, corresponding to the chosen effects and the mean, and their standard errors, and test them by the function t = regression coefficient/standard error [Table 11.5(c)]. In this way, the data can be analysed for significant effects, and also to test for non-zero mean.

11.5 ANALYSIS OF THE FITTED RESPONSE SURFACE

The response surface analysis should not end with even a well-fitted response equation. Apparently generally similar equations can give substantially different surfaces. Box, Hunter, and Hunter[11] showed that the hypothetical equations (11.17a–d) gave distinctly different surfaces [Figures 11.5(a)–(d)]:

$$y = b_0 + b_A x_A + b_B x_B + b_{AA} x_A^2 + b_{BB} x_B^2 + b_{AB} x_A x_B \quad (11.15)$$

$$y_a = 83.93 + 10.23 x_A + 5.59 x_B - 6.95 x_A^2 - 2.07 x_B^2 - 7.59 x_A x_B \quad (11.17a)$$

$$y_b = 82.71 + 8.80 x_A + 8.19 x_B - 6.95 x_A^2 - 2.07 x_B^2 - 7.59 x_A x_B \quad (11.17b)$$

$$y_c = 83.57 + 9.39x_A + 7.12x_B - 7.44x_A^2 - 3.71x_B^2 - 5.80x_Ax_B \quad (11.17c)$$

$$y_d = 84.29 + 11.06x_A + 4.05x_B - 6.46x_A^2 + 0.43x_B^2 - 9.38x_Ax_B \quad (11.17d)$$

The Figures 11.5(a)–(d) represent
(a) a stationary ridge (b) a rising ridge
(c) a simple maximum (d) a minimax*

We can get a general idea of the surface from the signs and magnitudes of the quadratic terms in the equations (see Section 10.2). The quadratic signs are negative in equations (11.17a, b, c), so each represents some form of maximum (or approach to it). The negative, positive signs in (11.17d) indicate a maximum with respect to A and a minimum with respect to B. A clearer idea can be obtained from the *canonical forms* of the equations, but graphics capability is nonetheless a highly desirable feature of software for RSM.

It may of course be that Figure 11.5(b) is part of a system of type 11.5 (c) or (d), and (a) may be part of an extreme ellipsoid. The rare minimax (d) reminds us that more than one optimum point may occur. The optimum is often called the *stationary point*, S. Differentiating equation 11.15 with respect to each variable, and setting $\delta y/\delta A = \delta y/\delta B = 0$, gives its co-ordinates as x_{As}, x_{Bs} in:

$$-b_A = 2b_{AA}x_{As} + b_{AB}x_{Bs}$$
$$-b_B = b_{AB}x_{As} + 2b_{BB}x_{Bs}$$

in which x_{As}, x_{Bs} are the only unknowns. The *predicted* (maximum or minimum) *response* y_s at S can be obtained by substituting x_{As}, x_{Bs} for x_A, x_B in equation (11.15).

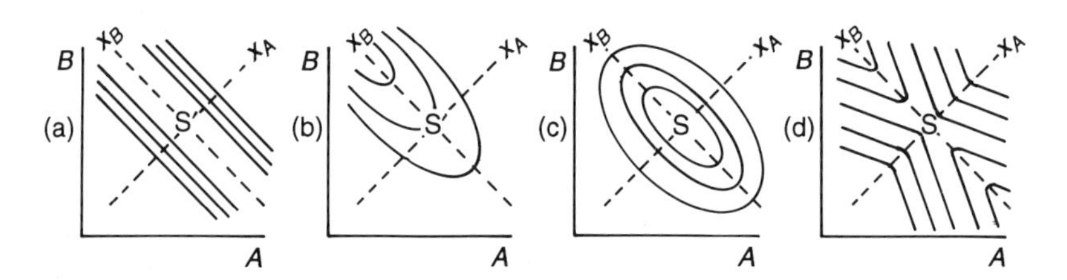

Figure 11.5 *Response surfaces from second-degree equations*

Re-writing a response surface equation by *canonical analysis* reveals its fundamental nature clearly. Canonical analysis consists of changing the origin of the plot from its original co-ordinates to the stationary point, and rotating the A, B axes until they correspond to the principal axes of the contours (see Figure 11.5). If, as in Figure 11.5 (a) and (b), the optimum is outside the experimental space, the new origin S′ is taken as near as convenient to the centre of the design.

This rearrangement results in the disappearance of the first-order terms from the response equations, and all but equation (11.17b) now take the form

$$y - y_s = B_{AA}X_A^2 + B_{BB}X_B^2 \quad (11.18)$$

The equations individually are

* Also known as a saddle or col.

$$y-87.69=-9.02X_A^2+0.00X_B^2 \quad (11.18a)$$

$$y-87.69=-9.02X_A^2-2.13X_B^2 \quad (11.18c)$$

$$y-87.69=-9.02X_A^2+2.13X_B^2 \quad (11.18d)$$

The rising ridge equation is now

$$y-87.69=-9.02X_A^2+2.97X_B \quad (11.18b)$$

This is the stationary ridge equation with an added first-order term in B, corresponding to the tilt along the canonical B axis.

We now see, for instance, from the negative signs in equation (11.18c), that moving from S in either the A or B direction will reduce the response, and that the decrease is more rapid in the A direction. The larger the coefficients, the more thorough must be the *control* of the variables if our results are to be acceptable.

Stationary ridges are fairly common, and should be exploited: they show that there is a series of combinations of levels for A and B, each of which will give optimum returns. The example shown in Figure 11.5(a) is the probably theoretical case where B_{BB}=zero. In practice, B_{BB} is close to zero, so the contours are slightly curved. In either case, they correspond to a marked AB interaction in 2^2 analysis. It may well be that changing the level of one of A, B is simpler, quicker, cheaper than for the other. Preferably, it will also be a reversible change, the extent of which is quickly and easily monitored. Hence, if A=pH, B=concentration of an expensive reagent, change pH. As another example, if the variables are catalyst concentration and temperature [and the interaction is positive: in Figure 11.5(a) it is negative] then increasing catalyst concentration gives a greater effect of temperature increase. For a given increase of yield, a smaller temperature increase is needed. Equally, increasing T means a smaller increase of [catalyst] for a given yield increase. The choice of which variable to change depends on the relative costs of heat and catalyst.

11.6 CHOICE OF OPERATING CONDITIONS TO ACHIEVE DESIRED SPECIFICATION

For simplicity, all the foregoing explanations have been in terms of one response. But often, we have to satisfy several criteria, and evaluate several responses as a consequence. It seems unlikely that we can optimize on all of them: we have to find the most efficient compromise condition. RSM contour plotting offers an obvious approach. Consider again the adjustment of catalyst concentration and temperature, where the necessary responses are yield, purity of product, and cost of operation. If the response contours for yield, purity, and cost of catalyst and heat were overlaid, they might look something like Figure 11.6.

If YY′ represents the minimum acceptable yield and II′ represents the maximum acceptable percentage of a damaging impurity, then the unhatched area will yield acceptable product, and the point P represents the minimum-cost process on these criteria.

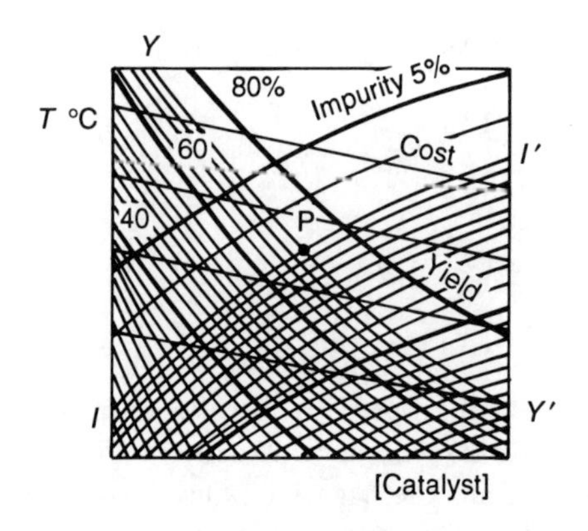

Figure 11.6 *Response curves for a chemical synthesis*

REFERENCES

1. C. Daniel, 'Applications of Statistics to Industrial Experimentation', Wiley, New York, 1979, pp. 61–63, 68–69.
2. G.E.P. Box and N.R. Draper, 'Empirical Model-building and Response Surfaces', Wiley, New York, 1987, Chapter 13.
3. Reference 2, pp. 481–489.
4. G.E.P. Box and K.B. Wilson, *J. Roy. Stat. Soc., Ser. B*, 1951, **13**, 1–45.
5. Reference 2.
6. G.E.P. Box, W.G. Hunter, and J.S. Hunter, 'Statistics for Experimenters', Wiley, New York, 1979, Chapter 15.
7. D.C. Montgomery, 'Design and Analysis of Experiments', 2nd Edn., Wiley, New York, 1984, Chapter 15.
8. E. Morgan, 'Chemometrics: Experimental Design', Wiley/ACOL, Chichester, 1991, Chapter 5.
9. N.R. Draper, *Technometrics*, 1985, **27**, 173–180.
10. N.R. Draper, *Technometrics*, 1982, **24**, 127–133.
11. Reference 6, pp. 527–530.

CHAPTER 12

Taguchi Methodology

Numerous statistical design methods exist which it has not been possible to include in this short text, but readers may expect me to comment briefly on a set of methods which have recently been enjoying some attention in the United Kingdom. The statistical ideas and methods employed by the engineering professor Genichi Taguchi, for improving the quality of products and processes, have been given much credit for the reliability and fitness-for-purpose of Japanese products. The concept of continuous quality improvement was actually introduced to Japanese industrialists in the early 1950s by the American statistician W.E. Deming, and Taguchi took up these ideas with enthusiasm, but apparently rather in isolation, for many years. By about 1980, the Japanese industrial success caused the importation of Taguchi's ideas into the USA, and then, from about 1985, into the UK.

In particular, Taguchi has emphasized the use of fractional factorial designs and other orthogonal arrays with the objectives of

(a) making products robust to environmental conditions;
(b) making products insensitive to component variation;
(c) minimizing bias and mean square error about specification values;
(d) reliability and life testing.

No-one could argue against these objectives of making robust, reliable, predictable, and adaptable products. Objective (c) epitomizes the application of statistical concepts to ensuring high-quality production, even of mass-produced goods: make one unit closely similar to another, and make them all close to what the customer has a right to expect. It is in some of Taguchi's statistical design and analysis concepts (and in the way they are sometimes put across by others) that there is cause for concern.

Taguchi has given much emphasis to 'lean' – *i.e.* 'saturated' – experiments. We saw in Section 4.1 that such designs involve much aliasing between factors and interactions. Although Taguchi often recommends the use of three-level designs, which we have seen are used when interactions are important, he argues against the need to consider two-factor interactions. He also recommends running one comprehensive experiment followed up by one confirmatory run of the supposedly best factor combination. This experiment can involve many replications of the basic design.

Cox long ago argued cogently against the 'comprehensive start' approach.[1] Firstly, there is the obvious consideration of simplicity: experiments with many factors and many runs are difficult to organize, even if the resources are available. Secondly, if the entire experiment is planned at the outset, one has to assume one knows which factors are most important, the ranges over which they should be studied, and in what metrics the factors and responses should be considered. Thirdly, it is more profitable in the long run to aim at scientific feedback (understanding the system) rather than empirical feedback (describing the effects of changing it). A sequence of relatively small designs is then called for, each suggested by previous ones, to explore the experimental space in terms of relevant factors. An example of possible consequences of Taguchi methods is as follows.

A UK manufacturer of car seats believed that seven variables affected the hardness of their cushions. As a consequence, an L_8 array (what we would call a 2^{7-4} fraction) was done.[2] Ten seats were made at each of the eight TCs, and the full run done a second time.[3] Of the seven variables chosen, isocyanate temperature and the ratio of polyol to isocyanate were found significant with regard to mean hardness, while the amount of hardener and shot time (determining the amount of injected material) were significant with regard to variability. A confirmatory run at the preferred settings of these variables (with the other variables set to values of lowest cost) gave cushions with mean hardness 221 N (target 220 N, previous mean 216 N) and standard deviation 3.6 N (previously 5.2 N). The product was reported as well within the customer's specification, and total cost reduction (direct production costs plus estimated savings from reduced customer complaints, defects replacement costs, *etc.*) was predicted at 17%.

So far, so good. But analysis of the TCs done reveals that the defining contrasts were *ABC*, *ADE*, *BDF*, *ABDG*; consequently each factor is aliased with three two-factor interactions, as in Table 12.1. Each factor is inevitably aliased with the other six, combined in pairs. The design is therefore of only resolution III, and there is much intertwining of concentration and temperature effects. Compare this situation with K.C. Peng's statement in his 1967 textbook:[4]

'The aliases system which describes the entangling situation should be carefully checked by the experimenter before he adopts a fractional factorial design.'

If a 2_{IV}^{7-3} design had been done (see, *e.g.*, Exercise 6.8), the factors would have been aliased only with three-factor interactions, although of course two-factor/two-factor aliasing would still occur. Keeping to the same total run for the moment, ten seats would have been produced at each of sixteen TCs, instead of twenty at each of eight, thus giving a more broadly based investigation. There would still have been 80 estimates of each factor with regard to the mean (and of up to eight chosen interactions as a bonus). There would still have been 40 estimates of each factor with regard to the variability (with interaction estimates again as a bonus). Here we have an example of 'the inefficient RIII designs produced by Taguchi followers when RIV could be established'[5] in what seems an excessively large programme of work. Running the 2^{7-3} design in duplicate (a total of 32 trials) would produce sixteen and eight estimates, respectively, and the runs could be replicated if necessary.

The analysis of variance with regard to the effects on the mean gave the

Table 12.1 *Two-factor aliases of chosen factors in seat hardness experiment*

Factor	*Two-factor aliases*
Isocyanate temperature*	(polyol temperature/isocyanate–polyol ratio) (amount of hardener/mould temperature) (ventilation control/shot time)
Isocyanate–polyol ratio*	(isocyanate temperature/polyol temperature) (amount of hardener/shot time) (mould temperature/ventilation control)
Amount of hardener†	(isocyanate temperature/mould temperature) (polyol temperature/ventilation control) (isocyanate–polyol ratio/shot time)
Shot time†	(amount of hardener/isocyanate–polyol ratio) (polyol temperature/mould temperature) (isocyanate temperature/ventilation control)

*Significant for mean.
†Significant for variability.

necessary one degree of freedom to each of four factors, and the remaining 155 to the error term. It is to all intents and purposes impossible not to find significance under such circumstances, whether there are significant effects or not. The extreme bias caused by even very much smaller pooling of variances is well known; it was because such methods almost guarantee spurious conclusions that Daniel introduced normal probability plotting. With regard to variability, a signal-to-noise ratio $SN_T = 10 \log_{10}(\bar{y}^2/s^2)$ was used. From the 2^{7-4} design, there could be only seven values for $\bar{y}$. Again giving $\phi = 1$ to each of four factors gave only $\phi = 3$ for error. Now, of course, significance will only be found if effects are relatively large. Taguchi's laudable aim is to maximize SN_T: to make the standard deviation as small as possible relative to the mean. But ST_T is ambiguous: a given value naturally represents different absolute variabilities dependent upon y.

Taguchi's ideas for making products insensitive to environmental or component variation are very important (as will be obvious to formulation chemists and those who have to consider the stability of products in store). Unfortunately, they can lead to very large designs. The principle is to combine a factorial in design (*inner*) factors with a factorial in *outer* factors, with each having the mean TC added. Thus, for three design factors and two outer factors, the total design has $(2^3+1)(2^2+1)=45$ TCs. If many factors have to be investigated, the designs become prohibitively large. For example, if the 2^{7-4} design outlined above was combined with three outer factors, there would be $(2^{7-4}+1)(2^3+1)=81$ TCs for a single run.

The strengths and weaknesses of Taguchi's ideas caused reactions in the United States which Gunter described as ranging from effusive praise to denunciation, and he concluded that there had been no scientific justification for the claimed superiority of those of Taguchi's methods which appear to be new.[6] There is much to value in Taguchi's advocacy of building in quality from the design stage (in contrast to seeking to achieve it by inspection and/or adjustment of the product) and his insistence on close adherence to specification (minimal systematic and random errors). Consequently, Box and his colleagues[7] have concluded that his quality engineering ideas are of great importance and should become part of the working knowledge of every engineer. But they have also concluded that many of his techniques of statistical design and analysis are often inefficient and unnecessarily complicated, and Greenfield[8] described them as limited, restricted, and misleading.

Perhaps I may end this book with a quotation from the Box, Bisgaard, and Fung paper just referred to:

'The challenge facing Western industry is enormous but not insuperable. Although the problems are multifaceted and no single solution exists, the use of statistical methods by everyone . . . with the enthusiastic support of an educated management, provides one of the most promising strategies for achieving excellence. Taguchi has certainly created enormous and much welcomed momentum for the use of experimental design in industry. However, . . . we feel that engineers working on designing quality into products and processes should be provided with the best *and* simplest statistical tools available.'

For engineers, read chemists (and chemical engineers, of course). I hope my book will prove an effective introduction to some of these tools.

REFERENCES

1. D.R. Cox, 'Planning of Experiments', Wiley, New York, 1958, pp. 95–96.
2. T. Shelley, *Eureka Innovative Engineering Design*, June, 1988.
3. G. Wilson, Coopers Lybrand seminar, April, 1989.
4. K.C. Peng, 'The Design and Analysis of Scientific Experiments', Addison-Wesley, Reading, Mass., 1967, p. 118.
5. J. Disney, *Appl. Stat.*, 1992, **41**, 229–231.
6. B. Gunter, *Qual. Assurance*, 1987, **13**, 81–87.
7. G.E.P. Box, S. Bisgaard, and C. Fung, *Qual. Reliability Eng. Int*, 1988, 124–131.
8. A.A. Greenfield, editorial, *Roy. Stat. Soc. News and Notes*, 1989, **15**, no. 8.

APPENDIX 1

Computer Programs for Experimental Design

Computer packages for statistical analysis have long been available, but interactive computer programs to aid experimental design have developed only since about 1985. The rapid proliferation of packages on offer has been due to the upsurge in the use of experimental design for the development of products and processes, itself caused by the need for quality improvement and the increased interest in Taguchi's philosophy. New and revised programs will doubtless continue to appear regularly, and some old ones will be discontinued. Consequently, there seems little point in attempting a 'state of the art' review, because the system is far from equilibrium; but some general notes may be of help. Those that follow are based on independent reviews published since 1987, except for programs marked *, for which information was gained from producers' literature or seminars.

An early but comprehensive review is that by Nachtsheim, who included detailed check-lists of capabilities which can be used to assess newer programs. Covering seventeen factorial and related designs such as are covered in this book, sixteen other less-used types, twenty-two aspects of statistical analysis, and thirteen aspects of process optimization, Nachtsheim reviewed two programs for optimization (Simplex-V and Ultramax) and nine for scientific feedback (Design-Ease, Aced, Cade, XStat, Statgraphics, Jass, Echip, Coed-RSM, and RS-Discover). Of these, I understand Jass is no longer supported by its producers. Programs that have appeared since the review include Quality and Productivity Improvement (QPI)* and Design-Expert from the USA, and Maximize* and Dex from the UK.

Software reviews began in the journal *Applied Statistics* in 1989, and at the time of writing had covered Statgraphics, X-Stat, Dex, Design-Ease and Design-Expert, and N, a program specifically for sample size and power calculations. Recent issues also include comments by manufacturers or agents. These reviews include details of the host language, minimum system requirement (and recommended requirement if this is greater), details of the system(s) on which the program was examined, and cost at time of review. Nachtsheim is weaker on some of these items, but more detailed on the actual system capabilities. Reviews also appear from time to time in, *e.g.*, *Chemistry and Industry* and *Computer Magazine*.

Nachtsheim identifies, for each program, what capabilities exist (including a sub-classification, possible but not simple to use) and gives an overall judgement on the ease of use of each program. He judged Design-Ease excellent in this regard; Cade, Echip, RS-Discover and Simplex V very good; XStat, Aced, Coed-RSM, Ultramax (and Jass) good. Statgraphics he did not examine: it is a relatively complete data analysis system with experimental design only a small part of the package. Accompanying documentation was classified as XStat, RS-Discover excellent; Design-Ease, Echip (and Jass) very good; Cade good; Aced and Coed-RSM satisfactory.

You may wish to make a check-list of your own requirements, initially based on the contents of this book, which you can compare with the capabilities of any package on offer. You will want, *e.g.*, numerical and graphical representations of means, standard deviation, standard error, confidence intervals, t- and F-tests; the design of 2^n and 2^{n-p} experiments, either by defining your own or choosing from a menu; sample size and power; blocking and randomizing of experiments; alias matrices; ANOVA; regression analysis and analysis of multi-variate designs; central composite designs; response surface methodology. If you contemplate a menu-driven package, you will want to know its contents, in terms of both design and analysis. If you contemplate a command-driven package, you will want to know how many commands there are, how brief they are, and so on. If you wish to design your own 2^n and 2^{n-p} experiments based on your own requirements set (and I hope you will) it seems you will currently have to look far for a package: Dex has an algorithm, but the 1991 review referenced below suggests that the total package is only in an early stage of development.

With regard to RSM, you will have numerous requirements, for without computers some of it would not be feasible. More recent packages provide them. You can choose up to, say, six factors, choose a design type from a menu list, and divide it into blocks appropriate to your purposes. You can transform factors and responses, choose an α value for a central composite design, choose the maximum degree of the polynomial you wish to fit the data to, and determine the regions in which two or more responses simultaneously meet specific requirements. The program will find the stationary point, predict the response there, and do a canonical analysis. Graphics capability allows for display of axis rotation and three-dimensional plotting.

The increasing scope of such software nonetheless does not mean that you can forget what you have learned from this or other books. For instance, you can find a program which includes an alias matrix, but only an incomplete one; or simpler software which does not include aliases at all. You may be able to divide a factorial into blocks for RSM, but not if you have specified the centroid (as you will have done). Some programs have unconventional terminology, so you need to know the principles to use them really effectively, and some feature less efficient designs as well as more efficient ones. Most important of all, the speed and precision of the computer complement, and should be complemented by, the slow, imprecise, and yet vast critical and inductive powers of the human mind.

Some programs, such as RS-Discover and Statgraphics, are completely menu driven; some, like QPI, have detailed manuals and instructional videotapes and are essentially user-command driven. The Design programs and QPI are modelled on the Box, Hunter, and Hunter textbook, and indeed the latter package can include a copy of the book along with its six videotapes. Some, like Maximize, have introductory tapes, training courses, user groups and consultant back-up. The costs of packages correspondingly vary widely; discounts for multiple copies, site licences, and educational use are commonplace.

	Program(s) reviewed
C.J. Nachtsheim, *J. Qual. Technol.*, 1987, **19**, 132–160	See text
R. Hunt, *Appl. Stat.*, 1989, **38**, 158–160	Statgraphics
B. Jones, *Appl. Stat.*, 1989, **38**, 392–395	XStat
D.O. Chanter, *Appl. Stat.*, 1989, **38**, 529–531	N
D. McAuliffe and C. Hamel, *Computer Magazine*, March, 1991, 86–92	Design-Ease
	Design-Expert
	Echip
	XStat
	Taguchi Analyst
B. Jones, *Appl. Stat.*, 1991, **40**, 491–493	Dex
K. Julian, *Chem. Ind.*, 2 March, 1992	Design-Ease
J. Disney, *Appl. Stat.*, 1992, **41**, 229–231	Design-Ease
	Design-Expert

APPENDIX 2

Percentage Points of the Normal Distribution

u_α is the value of the standardized Normal variate which has fractional probability α of being exceeded in a one-tailed test. The corresponding two-tailed probabilities are 2α, and all values become percentage probabilities when multiplied by 100.

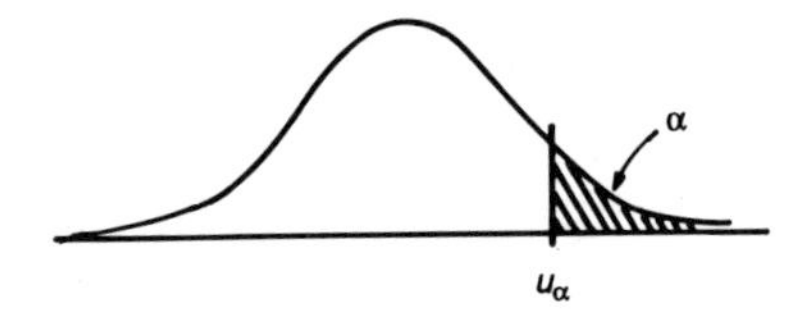

α	u_α	α	u_α	α	u_α
0.500	0.000	0.034	1.825	0.014	2.197
0.450	0.126	0.032	1.852	0.013	2.226
0.400	0.253	0.030	1.881	0.012	2.257
0.350	0.385	0.029	1.896	0.011	2.290
0.300	0.524	0.028	1.911	0.010	2.326
0.250	0.675	0.027	1.927	0.009	2.366
0.200	0.842	0.026	1.943	0.008	2.409
0.150	1.036	0.025	1.960	0.007	2.457
0.100	1.282	0.024	1.977	0.006	2.512
0.050	1.645	0.023	1.995	0.005	2.576
0.048	1.665	0.022	2.014	0.004	2.652
0.046	1.685	0.021	2.034	0.003	2.748
0.044	1.706	0.020	2.054	0.002	2.878
0.042	1.728	0.019	2.075	0.001	3.090
0.040	1.751	0.018	2.097	0.0005	3.291
0.038	1.774	0.017	2.120	0.00005	3.891
0.036	1.799	0.016	2.144	0.000005	4.417
		0.015	2.170		

APPENDIX 3

Percentage Points of the t-Distribution

The table gives the values of $t_{\alpha,\varphi}$ from the t distribution for ϕ degrees of freedom. The tabulation applies to one-tailed tests; for two-tailed tests, the column headings must be doubled to 2α.

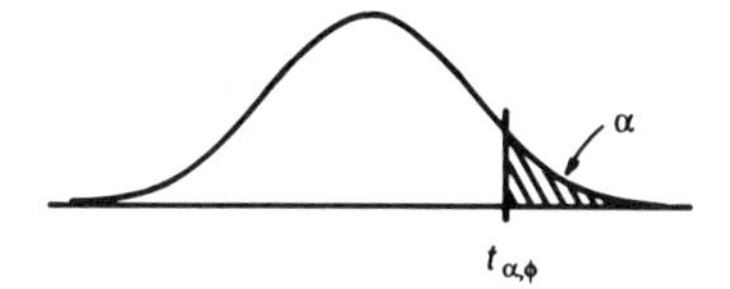

	$\alpha=$	0.10	0.05	0.025	0.01	0.005	0.001	0.0005
$\phi=$	1	3.078	6.314	12.706	31.821	63.657	318.31	636.62
	2	1.886	2.920	4.303	6.965	9.925	22.326	31.598
	3	1.638	2.353	3.182	4.541	5.841	10.213	12.924
	4	1.533	2.132	2.776	3.747	4.604	7.173	8.610
	5	1.476	2.015	2.571	3.365	4.032	5.893	6.869
	6	1.440	1.943	2.447	3.143	3.707	5.208	5.959
	7	1.415	1.895	2.365	2.998	3.499	4.785	5.408
	8	1.397	1.860	2.306	2.896	3.355	4.501	5.041
	9	1.383	1.833	2.262	2.821	3.250	4.297	4.781
	10	1.372	1.812	2.228	2.764	3.169	4.144	4.587
	11	1.363	1.796	2.201	2.718	3.106	4.025	4.437
	12	1.356	1.782	2.179	2.681	3.055	3.930	4.318
	13	1.350	1.771	2.160	2.650	3.012	3.852	4.221
	14	1.345	1.761	2.145	2.624	2.977	3.787	4.140
	15	1.341	1.753	2.131	2.602	2.947	3.733	4.073
	16	1.337	1.746	2.120	2.583	2.921	3.686	4.015
	17	1.333	1.740	2.110	2.567	2.898	3.646	3.965
	18	1.330	1.734	2.101	2.552	2.878	3.610	3.922
	19	1.328	1.729	2.093	2.539	2.861	3.579	3.883
	20	1.325	1.725	2.086	2.528	2.845	3.552	3.850
	21	1.323	1.721	2.080	2.518	2.831	3.527	3.819
	22	1.321	1.717	2.074	2.508	2.819	3.505	3.792
	23	1.319	1.714	2.069	2.500	2.807	3.485	3.767
	24	1.318	1.711	2.064	2.492	2.797	3.467	3.745
	25	1.316	1.708	2.060	2.485	2.787	3.450	3.725

$\alpha=$	0.10	0.05	0.025	0.01	0.005	0.001	0.0005
26	1.315	1.706	2.056	2.479	2.779	3.435	3.707
27	1.314	1.703	2.052	2.473	2.771	3.421	3.690
28	1.313	1.701	2.048	2.467	2.763	3.408	3.674
29	1.311	1.699	2.045	2.462	2.756	3.396	3.659
30	1.310	1.697	2.042	2.457	2.750	3.385	3.646
40	1.303	1.684	2.021	2.423	2.704	3.307	3.551
60	1.296	1.671	2.000	2.390	2.660	3.232	3.460
120	1.289	1.658	1.980	2.358	2.617	3.160	3.373
∞	1.282	1.645	1.960	2.326	2.576	3.090	3.291

This table is taken from Table III of Fisher and Yates: 'Statistical Tables for Biological, Agricultural and Medical Research', published by Longman Group UK Ltd, 1974; and from Table 12, 'Biometrika Tables for Statisticians', Vol. 1, 3rd Edn., 1966, by permission of the Biometrika Trustees.

APPENDIX 4

Percentage Points of the *F*-Distribution

The table overleaf gives the values of $F_{\alpha;\varphi 1,\varphi 2}$ from the F distribution with ϕ_1 degrees of freedom in the numerator and ϕ_2 degrees of freedom in the denominator. For each pair of values, ϕ_1 and ϕ_2, $F_{\alpha;\varphi 1,\varphi 2}$ is tabulated for $\alpha = 0.05$, 0.025, 0.01, 0.001, with 0.025 values in brackets, for one-tailed tests. The same F-values correspond to 2α for two-tailed tests.

This table is taken from Table V of Fisher and Yates: 'Statistical Tables for Biological, Agricultural and Medical Research', published by Longman Group UK Ltd, 1974; and from Table 18, 'Biometrika Tables for Statisticians,' Vol. 1, 3rd Edn., 1966, by permission of the Biometrika Trustees.

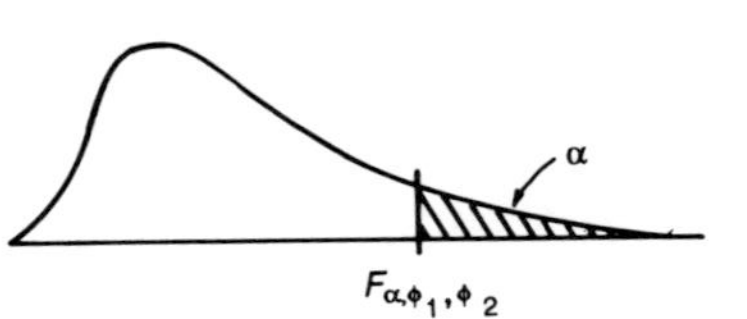

ϕ_2	ϕ_1 1	2	3	4	5	6	7	8	10	12	24	∞
1	161.4	199.5	215.7	224.6	230.2	234.0	236.8	238.9	241.9	243.9	249.0	254.3
	(648)	(800)	(864)	(900)	(922)	(937)	(948)	(957)	(969)	(977)	(997)	1018)
	4052	5000	5403	5625	5764	5859	5928	5918	6056	6106	6235	6366
	4053*	5000*	5404*	5625*	5764*	5859*	5929*	5981*	6056*	6107*	6235*	6366*
2	18.5	19.0	19.2	19.2	19.3	19.3	194.	19.4	19.4	19.4	19.5	19.5
	(38.5)	(39.0)	(39.2)	(39.2)	(39.3)	(39.3)	(39.4)	(39.4)	(39.4)	(39.4)	(39.5)	(39.5)
	98.5	99.0	99.2	99.2	99.3	99.3	99.4	99.4	99.4	99.4	99.5	99.5
	998.5	999.0	999.2	999.2	999.3	999.3	999.4	999.4	999.4	999.4	999.5	999.5
3	10.13	9.55	9.28	9.12	9.01	8.94	8.89	8.85	8.79	8.74	8.64	8.53
	(17.4)	(16.0)	(15.4)	(15.1)	(14.9)	(14.7)	(14.6)	(14.5)	(14.4)	(14.3)	(14.1)	(13.9)
	34.1	30.8	29.5	28.7	28.2	27.9	27.7	27.5	27.2	27.1	26.6	26.1
	167.0	148.5	141.1	137.1	134.6	132.8	131.5	130.6	129.2	128.3	125.9	123.5
4	7.71	6.94	6.59	6.39	6.26	6.16	6.09	6.04	5.96	5.91	5.77	5.63
	(12.22)	(10.65)	(9.98)	(9.60)	(9.36)	(9.20)	(9.07)	(8.98)	(8.84)	(8.75)	(8.51)	(8.26)
	21.2	18.0	16.7	16.0	15.5	15.2	15.0	14.8	14.5	14.4	13.9	13.5
	74.14	61.25	56.18	53.44	51.71	50.53	49.66	49.00	48.05	47.41	45.77	44.05
5	6.61	5.79	5.41	5.19	5.05	4.95	4.88	4.82	4.74	4.68	4.53	4.36
	(10.01)	(8.43)	(7.76)	(7.39)	(7.15)	(6.98)	(6.85)	(6.76)	(6.62)	(6.52)	(6.28)	(6.02)
	16.26	13.27	12.06	11.39	10.97	10.67	10.46	10.29	10.05	9.89	9.47	9.02
	47.18	37.12	33.20	31.09	29.75	28.83	28.16	27.65	26.92	26.42	25.14	23.79
6	5.99	5.14	4.76	4.53	4.39	4.28	4.21	4.15	4.06	4.00	3.84	3.67
	(8.81)	(7.26)	(6.60)	(6.23)	(5.99)	(5.82)	(5.70)	(5.60)	(5.46)	(5.37)	(5.12)	(4.85)
	13.74	10.92	9.78	9.15	8.75	8.47	8.26	8.10	7.87	7.72	7.31	6.88
	35.51	27.00	23.70	21.92	20.80	20.03	19.46	19.03	18.41	17.99	16.90	15.75

7	5.59	4.74	4.35	4.12	3.97	3.87	3.79	3.73	3.64	3.57	3.41	3.23
	(8.07)	(6.54)	(5.89)	(5.52)	(5.29)	(5.12)	(4.99)	(4.90)	(4.76)	(4.67)	(4.42)	(4.14)
	12.25	9.55	8.45	7.85	7.46	7.19	6.99	6.84	6.62	6.47	6.07	5.65
	29.25	21.69	18.77	17.20	16.21	15.52	15.02	14.63	14.08	13.71	12.73	11.70
8	5.32	4.46	4.07	3.84	3.69	3.58	3.50	3.44	3.35	3.28	3.12	2.93
	(7.57)	(6.06)	(5.42)	(5.05)	(4.82)	(4.65)	(4.53)	(4.43)	(4.30)	(4.20)	(3.95)	(3.67)
	11.26	8.65	7.59	7.01	6.63	6.37	6.18	6.03	5.81	5.67	5.28	4.86
	25.42	18.49	15.83	14.39	13.48	12.86	12.40	12.05	11.54	11.19	10.30	9.34
9	5.12	4.26	3.86	3.63	3.48	3.37	3.29	3.23	3.14	3.07	2.90	2.71
	(7.21)	(5.71)	(5.08)	(4.72)	(4.48)	(4.32)	(4.20)	(4.10)	(3.96)	(3.87)	(3.61)	(3.33)
	10.56	8.02	6.99	6.42	6.06	5.80	5.61	5.47	5.26	5.11	4.73	4.31
	22.86	16.39	13.90	12.56	11.71	11.13	10.69	10.37	9.87	9.57	8.72	7.81
10	4.96	4.10	3.71	3.48	3.33	3.22	3.14	3.07	2.98	2.91	2.74	2.54
	(6.94)	(5.46)	(4.83)	(4.47)	(4.24)	(4.07)	(3.95)	(3.85)	(3.72)	(3.62)	(3.37)	(3.08)
	10.04	7.56	6.55	5.99	5.64	5.39	5.20	5.06	4.85	4.71	4.33	3.91
	21.04	14.91	12.55	11.28	10.48	9.93	9.52	9.20	8.74	8.44	7.64	6.76
11	4.84	3.98	3.59	3.36	3.20	3.09	3.01	2.95	2.85	2.79	2.61	2.40
	(6.72)	(5.26)	(4.63)	(4.28)	(4.04)	(3.88)	(3.76)	(3.66)	(3.53)	(3.43)	(3.17)	(2.88)
	9.65	7.21	6.22	5.67	5.32	5.07	4.89	4.74	4.54	4.40	4.02	3.60
	19.69	13.81	11.56	10.35	9.58	9.05	8.66	8.35	7.92	7.63	6.85	6.00
12	4.75	3.89	3.49	3.26	3.11	3.00	2.91	2.85	2.75	2.69	2.51	2.30
	(6.55)	(5.10)	(4.47)	(4.12)	(3.89)	(3.73)	(3.61)	(3.51)	(3.37)	(3.28)	(3.02)	(2.72)
	9.33	6.93	5.95	5.41	5.06	4.82	4.64	4.50	4.30	4.16	3.78	3.36
	18.64	12.97	10.80	9.63	8.89	8.38	8.00	7.71	7.29	7.00	6.25	5.42
13	4.67	3.81	3.41	3.18	3.03	2.92	2.83	2.77	2.67	2.60	2.42	2.21
	(6.41)	(4.97)	(4.35)	(4.00)	(3.77)	(3.60)	(3.48)	(3.39)	(3.25)	(3.15)	(2.89)	(2.60)
	9.07	6.70	5.74	5.21	4.86	4.62	4.44	4.30	4.10	3.96	3.59	3.17
	17.82	12.31	10.21	9.07	8.35	7.86	7.49	7.21	6.80	6.52	5.78	4.97

*Entries marked thus must be multiplied by 100.

ϕ_2 \ ϕ_1	1	2	3	4	5	6	7	8	10	12	24	∞
14	4.60	3.74	3.34	3.11	2.96	2.85	2.76	2.70	2.60	2.53	2.35	2.13
	(6.30)	(4.86)	(4.24)	(3.89)	(3.66)	(3.50)	(3.38)	(3.29)	(3.15)	(3.05)	(2.79)	(2.49)
	8.86	6.51	5.56	5.04	4.70	4.46	4.28	4.14	3.94	3.80	3.43	3.00
	17.14	11.78	9.73	8.62	7.92	7.44	7.08	6.80	6.40	6.13	5.41	4.60
18	4.41	3.55	3.16	2.93	2.77	2.66	2.58	2.51	2.41	2.34	2.15	1.92
	(5.98)	(4.56)	(3.95)	(3.61)	(3.38)	(3.22)	(3.10)	(3.01)	(2.87)	(2.77)	(2.50)	(2.19)
	8.29	6.01	5.09	4.58	4.25	4.01	3.84	3.71	3.51	3.37	3.00	2.57
	15.38	10.39	8.49	7.46	6.81	6.35	6.02	5.76	5.39	5.13	4.45	3.67
22	4.30	3.44	3.05	2.82	2.66	2.55	2.46	2.40	2.30	2.23	2.03	1.78
	(5.79)	(4.38)	(3.78)	(3.44)	(3.22)	(3.05)	(2.93)	(2.84)	(2.70)	(2.60)	(2.33)	(2.00)
	7.95	5.72	4.82	4.31	3.99	3.76	3.59	3.45	3.26	3.12	2.75	2.31
	14.38	9.61	7.80	6.81	6.19	5.76	5.44	5.19	4.83	4.58	3.92	3.15
26	4.23	3.37	2.98	2.74	2.59	2.47	2.39	2.32	2.22	2.15	1.95	1.69
	(5.66)	(4.27)	(3.67)	(3.33)	(3.10)	(2.94)	(2.82)	(2.73)	(2.59)	(2.49)	(2.22)	(1.88)
	7.72	5.53	4.64	4.14	3.82	3.59	3.42	3.29	3.09	2.96	2.58	2.13
	13.74	9.12	7.36	6.41	5.80	5.38	5.07	4.83	4.48	4.24	3.59	2.82
30	4.17	3.32	2.92	2.69	2.53	2.42	2.33	2.27	2.16	2.09	1.89	1.62
	(5.57)	(4.18)	(3.59)	(3.25)	(3.03)	(2.87)	(2.75)	(2.65)	(2.51)	(2.41)	(2.14)	(1.79)
	7.56	5.39	4.51	4.02	3.70	3.47	3.30	3.17	2.98	2.84	2.47	2.01
	13.29	8.77	7.05	6.12	5.53	5.12	4.82	4.58	4.24	4.00	3.36	2.59
40	4.08	3.23	2.84	2.61	2.45	2.34	2.25	2.18	2.08	2.00	1.79	1.51
	(5.42)	(4.05)	(3.46)	(3.13)	(2.90)	(2.74)	(2.62)	(2.53)	(2.39)	(2.29)	(2.01)	(1.64)
	7.31	5.18	4.31	3.83	3.51	3.29	3.12	2.99	2.80	2.66	2.29	1.80
	12.61	8.25	6.59	5.70	5.13	4.73	4.44	4.21	3.87	3.64	3.01	2.23
∞	3.84	3.00	2.60	2.37	2.21	2.10	2.01	1.94	1.83	1.75	1.52	1.00
	(5.02)	(3.69)	(3.12)	(2.79)	(2.57)	(2.41)	(2.29)	(2.19)	(2.05)	(1.94)	(1.64)	(1.00)
	6.63	4.61	3.78	3.32	3.02	2.80	2.64	2.51	2.32	2.18	1.79	1.00
	10.83	6.91	5.42	4.62	4.10	3.74	3.47	3.27	2.96	2.74	2.13	1.00

Subject Index